Lecture Notes in Biomathematics

Managing Editor: S. Levin

20

J. M. Cushing

Integrodifferential Equations and Delay Models in Population Dynamics

Springer-Verlag
Berlin Heidelberg New York

Lecture Notes in Biomathematics

Vol. 1: P. Waltman, Deterministic Threshold Models in the Theory of Epidemics. 101 pages. 1974.

Vol. 2: Mathematical Problems in Biology, Victoria Conference 1973. Edited by P. va den Driessche. VI, 280 pages. 1974.

Vol. 3: D. Ludwig, Stochastic Population Theories. VI, 108 pages. 1974.

Vol. 4: Physics and Mathematics of the Nervous System. Edited by M. Conrad, W. Gü tinger, and M. Dal Cin. XI, 584 pages. 1974.

Vol. 5: Mathematical Analysis of Decision Problems in Ecology. Proceedings 197 Edited by A. Charnes and W. R. Lynn. VIII, 421 pages. 1975.

Vol. 6: H. T. Banks, Modeling and Control in the Biomedical Sciences. V, 114 pages. 197

Vol. 7: M. C. Mackey, Ion Transport through Biological Membranes, An Integrate Theoretical Approach. IX, 240 pages. 1975.

Vol. 8: C. DeLisi, Antigen Antibody Interactions. IV, 142 pages. 1976.

Vol. 9: N. Dubin, A Stochastic Model for Immunological Feedback in Carcinogenesi Analysis and Approximations. XIII, 163 pages. 1976.

Vol. 10: J. J. Tyson, The Belousov-Zhabotinskii Reaktion. IX, 128 pages. 1976.

Vol. 11: Mathematical Models in Medicine. Workshop 1976. Edited by J. Berger, W. Bü ler, R. Repges, and P. Tautu. XII, 281 pages. 1976.

Vol. 12: A. V. Holden, Models of the Stochastic Activity of Neurones. VII, 368 pages. 197

Vol. 13: Mathematical Models in Biological Discovery. Edited by D. L. Solomon an C. Walter. VI, 240 pages. 1977.

Vol. 14: L. M. Ricciardi, Diffusion Processes and Related Topics in Biology. VI, 200 page 1977.

Vol. 15: Th. Nagylaki, Selection in One- and Two-Locus Systems. VIII, 208 pages. 197

Vol. 16: G. Sampath, S. K. Srinivasan, Stochastic Models for Spike Trains of Sing Neurons. VIII, 188 pages. 1977.

Vol. 17: T. Maruyama, Stochastic Problems in Population Genetics. VIII, 245 pages. 197

Vol. 18: Mathematics and the Life Sciences. Proceedings 1975. Edited by D. E. Matthew VII, 385 pages. 1977.

Vol. 19: Measuring Selection in Natural Populations. Edited by F. B. Christiansen an T. M. Fenchel. XXXI, 564 pages. 1977.

Vol. 20: J. M. Cushing, Integrodifferential Equations and Delay Models in Populatio Dynamics. VI, 196 pages. 1977.

Lecture Notes in Biomathematics

Managing Editor: S. Levin

20

J. M. Cushing

Integrodifferential Equations and Delay Models in Population Dynamics

Springer-Verlag
Berlin Heidelberg New York 1977

Author
Jim M. Cushing
Department of Mathematics and
Program in Applied Mathematics
University of Arizona
Tucson, Arizona 85721/USA

AMS Subject Classifications (1970): 45J05, 92-02, 92A15

ISBN 3-540-08449-5 Springer-Verlag Berlin Heidelberg New York
ISBN 0-387-08449-5 Springer-Verlag New York Heidelberg Berlin

Printing and binding: Beltz Offsetdruck, Hemsbach/Bergstr.
2145/3140-543210

PREFACE

These notes are, for the most part, the result of a course I taught at the University of Arizona during the Spring of 1977. Their main purpose is to investigate the effect that delays (of Volterra integral type) have when placed in the differential models of mathematical ecology, as far as stability of equilibria and the nature of oscillations of species densities are concerned. A secondary purpose of the course out of which they evolved was to give students an (at least elementary) introduction to some mathematical modeling in ecology as well as to some purely mathematical subjects, such as stability theory for integrodifferential systems, bifurcation theory, and some simple topics in perturbation theory. The choice of topics of course reflects my personal interests; and while these notes were not meant to exhaust the topics covered, I think they and the list of references come close to covering the literature to date, as far as integrodifferential models in ecology are concerned.

I would like to thank the students who took the course and consequently gave me the opportunity and stimulus to organize these notes. Special thanks go to Professor Paul Fife and Dr. George Swan who also sat in the course and were quite helpful with their comments and observations. Also deserving thanks are Professor Robert O'Malley and Ms. Louise C. Fields of the Applied Mathematics Program here at the University of Arizona. Ms. Fields did an outstandingly efficient and accurate typing of the manuscript.

J.M. Cushing

TABLE OF CONTENTS

CHAPTER 1. INTRODUCTORY REMARKS

Many, if not most differential models which are used in an attempt to describe dynamics of a population of a single species or several interacting species are mulated by prescribing in some appropriate manner the instantaneous per unit wth rate of each species:

$$N_i'/N_i = f_i, \quad i = 1, \ldots, n, \quad n \geq 1. \tag{1.1}$$

se models assume of course that the population sizes or densities are large and least in simpler cases ignore many complicating factors such as those due to age ucture, spacial distribution and diffusion, sexual categories, the presence of e delays and others. Our purpose in these notes is to consider such models n an attempt is made to account for time delays in the response of growth rates changes in population sizes.

In (1.1) the function $N_i = N_i(t)$ is some appropriate measure of the size or sity of the ith species (e.g. biomass) as a function of time t and $= dN_i/dt$ is its rate of change. The essential part of the modeling procedure s of course in the description of the expressions f_i. If for example f_i is onstant independent of time t and also of every N_j, $1 \leq j \leq n$ then one obns exponential (Malthusian) growth or decay for N_i depending on whether > 0 or $f_i < 0$. This mathematically uninteresting and biologically unrealistic ult can be modified in order to obtain more realistic models by assuming that depends upon the populations sizes N_j (or at least some of them) at time t: $= f_i(t,N_1(t), \ldots, N_n(t))$. This leads to a system of ordinary differential ations (of Kolomogorov type) to which the general theory and techniques of this t subject are applicable. These equations possess an extensive literature, both this general setting and with regard to more specific special cases derived for one type of interacting community or another.

Our purpose here is not to consider per se the differential model just described, but to consider models in which the rates f_i depend more generally on the population sizes $N_j(s)$, $s \leq t$, at past times s previous to t. Such an assumption leads to functional differential equations which, although more difficult to study in general, also fall within the purview of a well-developed general theory. Even so, such models have received little attention in the literature in comparison to the corresponding models without delays, despite the fact that Volterra introduced and studied some delay models in his work on population dynamics as early as the 1920's and 1930's (see Scudo's (1971) survey paper for a complete bibliography of Volterra's work). Inasmuch as it is generally recognized that delays play a significant roll in population dynamics (and biology in general) models with delays have been more frequently proposed and studied in recent years as is evidenced by an expanding literature on the subject.

Time delays in the growth dynamics of a population or of several interacting species can arise from a great variety of causes and are undoubtedly always present to some extent. For example, it is widely recognized that the population growth rate of a single (isolated) species is decreased with increased density of the species, a fact which often leads investigators to use the famous logistic or Pearl-Verhulst equation as a model. At times, however, it is difficult to fit data to a logistic curve even for simple, controlled laboratory experiments and one major problem as pointed out by F. Smith (1963) is due to time delays in the growth rate response (in the form of changes in natality and mortality) to density changes. Thus, even for such relatively simple organisms as bacteria continuous culture techniques have been used in order to eliminate problems associated with delays in order to study their population growth characteristics (F. Smith (1954)). Oscillations are often observed in such laboratory experiments and attributed to time delayed responses (see Nicholson (1957), Caperon (1969), May et al. (1974),

. M. Smith (1974) and the references cited in these works).

One frequently considered mechanism which introduces delays into the dynamics f population growth is that of age structure. Most of Volterra's original work n population dynamics in which delays were considered is in fact based on the onsideration of some aspect of age structure in the population or within individal species. The response of the growth rate of a population to increased populaion densities may for example be delayed because of a maturation period (e.g. due o a larval stage in insect populations, see Nicholson (1957, J. M. Smith (1974)) r a gestation period (e.g. of a predator after contact with prey, see Volterra 1931), Scudo (1971)). Other delay mechanisms which have been mentioned in the iterature include feeding times and hunger coefficients in predator-prey interctions (Caswell (1972)) and replenishment or regeneration time for resources (e.g. f vegetation for grazing herbivores, May (1973), May et al. (1974)). One can asily imagine other causes of delays in population dynamics on various time cales: those caused by food storage of predators or gatherers, reaction times, hreshold levels, etc.

We will not attempt here to study functional differential equations of the orm (1.1) in any great generality. Instead we will consider (1.1) under various estrictions on f_i which although specialized will nonetheless be general enough o include most of the models which have been considered in the literature either n delay or nondelay form. Our main concern will be with the asymptotic behavior as $t \to +\infty$) of solutions or more specifically with such topics as the stability r instability of equilibria and with the oscillatory nature of solutions. Moreover, our analysis will be for the most part local in nature, as it also freuently is for nondelay models. It is true that global phenomena are perhaps the ore important in the ecologist's view and that the often local (i.e. near equilibium) nature of the mathematician's analysis is frequently criticized. Nonetheless, ocal analysis near equilibria constitutes a first step in the analysis and understanding of differential equations. (One reason, if not the main reason being certainly that the mathematical techniques needed to accomplish such local analysis are more readily available.)

We will assume throughout that the delays which appear in (1.1) are of the Volterra integral form

$$\int_{-\infty}^{t} g(N(s))k(t - s)ds = \int_{0}^{\infty} g(N(t - s)k(s)ds \tag{1.2}$$

except that occasionally these integrals will be replaced by more general Stieltjes integrals. This type of functional expression was used by Volterra (1931). Its appearance in (1.1) causes the growth rates at time t to be effected by population sizes at (possibly all) previous times $s \leq t$ in a manner determined by the function g and distributed (or weighted) in the past by the <u>delay kernel</u> $k(s)$. To avoid various technical concepts and difficulties we will always assume $k(s)$ to be at least piece-wise continuous and satisfy $k(s)\varepsilon L^1$, that is

$$|k|_1 := \int_0^{\infty} |k(s)|ds < +\infty. \tag{1.3}$$

If $k\varepsilon L^1$ and in addition $k(s) \geq 0$ for all $s \geq 0$ we write $k\varepsilon L^1_+$. If delay effects have a maximum range of effectiveness in the sense that population densities at more than t_0 $(0 \leq t_0 < +\infty)$ time units into the past can have no influence on growth rates then $k(s) \equiv 0$ for $s \leq t_0$ and (1.3) is automatically fulfilled. However, we will not usually make this seemingly simplifying assumption (which is perhaps quite biologically reasonable) since it usually makes the mathematical details of specific models more difficult while delay kernels expressible in terms of elementary functions (such as polynomials multiplied by exponentials) without such so-called "compact support" are certainly qualitatively just as valid.

Strictly speaking we have ruled out differential-difference equations as dels in which delay effects appear as time lag expressions of the form t - T), $T > 0$. We could, and in fact will upon occasion, include time lag dels in our analysis by replacing $k(s)ds$ by $dh(s)$ so that the integrals .2) become Stieltjes integrals. Then by using step functions $u_T(s)$ for tegrators $h(s)$ we could obtain from (1.2) lag terms of the form $g(N(t - T))$. cause the "continuously distributed" delay models with $k(s)ds$ are obviously re realistic and, as it turns out, more amenable to our analysis we prefer to udy them in place of the more general models with Stieltjes integrals. There also some experimental evidence which indicates that such "continuously dis-ibuted" delays are more accurate than those with instantaneous time lags (see peron (1969)).

Although much of our work below will be done for general delay kernels, two neric types will be kept in mind throughout and frequently used in illustra-ons and specific model examples. These are $k(t) = T^{-1} \exp(-t/T)$ and t) $= T^{-2} t \exp(-t/T)$ for $T > 0$. The first kernel qualitatively represents a eak" delay in the sense that the maximum (weighted) response of the growth te is to current population density while past densities have (exponentially) creasing influence. The second kernel on the other hand qualitatively repre-nts the case when the maximum influence on growth rate response at any time t due to population density at the previous time $t - T$ (i.e. the maximum of is kernel occurs at $t = T$, see FIGURE 1.1). This latter kernel is a "con-nuously distributed" delay analog of a single time lag of length T. Both of ese kernels have been normalized so that $|k|_1 = 1$.

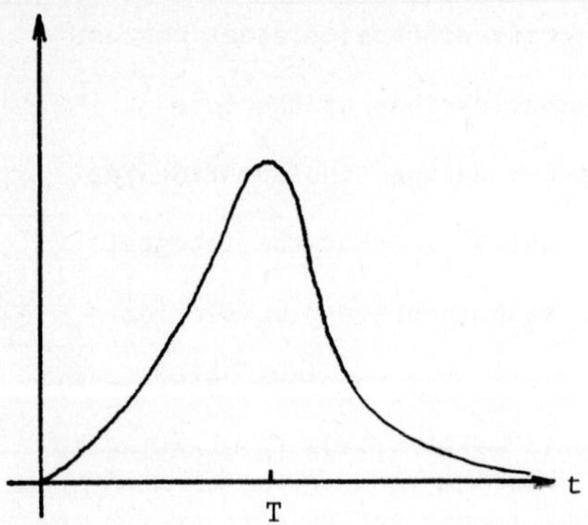

"strong" generic delay kernel

$k(t) = T^{-2}t \exp(-t/T)$

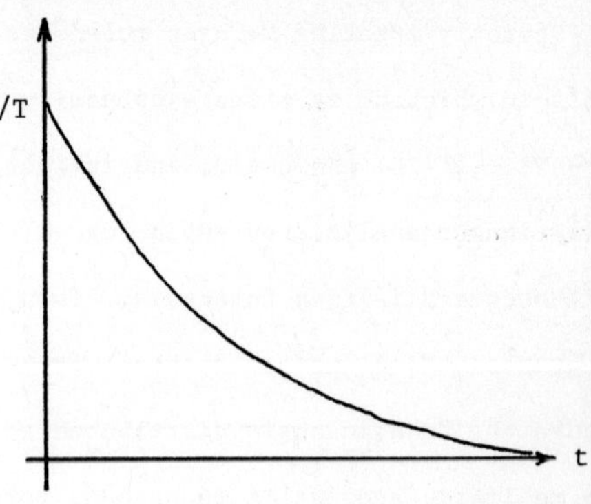

"weak" generic delay kernel

$k(t) = T^{-1} \exp(-t/T)$

FIGURE 1.1

One can also consider linear combinations of these kernels in order to obtain delay kernels in which some instantaneous effect on growth rate response is present, but the maximum effect is delayed. For example, such is the case for $k(t) = (aT^{-1} + bT^{-2}t) \exp(-t/T)$, $a + b = 1$, $b > a \geq 0$ for which $k(0) = aT^{-1} > 0$ and the maximum occurs at $t = T(b - a)/b$.

Let t_0, $-\infty < t_0 < +\infty$ be a fixed reference time. By a <u>solution</u> of (1.1) on an interval $t_0 < t < t_0 + \delta$, $0 < \delta \leq +\infty$ we mean a set of functions $N_i(t)$ which are (for simplicity) piece-wise continuous for all t: $-\infty < t < t_0 + \delta$, are differentiable for $t_0 < t < t_0 + \delta$ and reduce (1.1) to an identity on $t_0 < t < t_0 + \delta$. Following Volterra (1931) we can obtain general existence, uniqueness, extendibility and continuity (with respect to parameters and initial data) results by setting $N_i(t) \equiv N_i^0(t)$, $-\infty < t \leq t_0$ for given piece-wise continuous initial functions $N_i^0(t)$ and thereby reducing (1.1) to a Volterra integro-differential system with initial values at $t = t_0$ to which standard theorems ply (see Miller (1971)). It would be easy to write down such theorems by re-iring any of the familiar smoothness restrictions (e.g. a Lipschitz condition) f_i in (1.1). However, since such fundamental results are not our concern re we will refrain from doing this.

Certainly it is only solutions of (1.1) satisfying $N_i(t) > 0$ for all t ich are of interest as far as their potential use in ecological applications is ncerned. By a *positive solution* of (1.1) on $t_0 < t < t_0 + \delta$ we mean a solu-on as defined above for which $N_i(t) > 0$ for all $1 \leq i \leq n$ and $< t < t_0 + \delta$. By an *equilibrium* we mean a solution $N_i(t) \equiv e_i$, $-\infty < t < +\infty$ r constants e_i.

THEOREM 1.1 (Positivity Theorem) *If* $N_i(t)$ *is a solution of* (1.1) *on* $< t < t_0 + \delta$, $0 < \delta \leq +\infty$ *for which* $N_i^0(t) > 0$ *for all* $t \leq t_0$ *and* $1 \leq i \leq n$ *en* $N_i(t)$ *is a positive solution on* $t_0 < t < t_0 + \delta$.

Proof. Let f_i^* denote f_i evaluated at $N_i(t)$, $-\infty < t < t_0 + \delta$ and con-der the system $N_i^{*\prime}/N_i^* = f_i^*$ for an unknown N_i^*. Clearly $N_i^* \equiv N_i$ solves this near problem and hence $N_i(t) = N_i(t_0)\exp(f_i^*)$ for $t_0 < t < t_0 + \delta$ from which llows $N_i(t) > 0$ on $t_0 < t < t_0 + \delta$. □

Thus a solution of (1.1) which "starts positive, stays positive."

CHAPTER 2. SOME PRELIMINARY REMARKS ON STABILITY

2.1 Linearization. We wish to concern ourselves here only with the simplest and most commonly considered notion of (Liapunov) stability for differential systems. Our approach will be the standard one of the linearization of the model about a solution (usually an equilibrium) whose stability or instability is of interest. This of course yields stability results which are in general only local near the solution being considered. Although such local stability results are sometimes criticized as being too restrictive and inappropriate in general for the study of what should be a broader concept of "ecological stability," we will justify this approach by saying that such a study surely contributes at least a first step towards understanding the qualitative nature of the models considered here.

Without any loss in generality we take the reference time t_0 in the definitions of Chapter 1 from now on to be $t_0 = 0$. Let R^n denote n-dimensional Euclidean space and for simplicity let $R = R^1$ denote the real line. Let R^+ denote the positive reals and let R^- denote the nonpositive reals. If $\overline{N}_i$ is a solution of (1.1) on R then it is called stable if given any $\varepsilon > 0$ there exists a corresponding $\delta = \delta(\varepsilon) > 0$ such that $|N_i^0(t) - \overline{N}_i(t)| \leq \delta$ for all $t \varepsilon R^-$ and all i implies that any solution of (1.1) satisfying $N_i(t) = N_i^0(t)$ for $t \varepsilon R^-$ exists and satisfies $|N_i(t) - \overline{N}_i(t)| \leq \varepsilon$ for all $t \varepsilon R^+$ and all i. If in addition there exists a constant $\delta_0 > 0$ such that $|N_i^0(t) - \overline{N}_i(t)| \leq \delta_0$ on R^- for all i implies that $|N_i(t) - \overline{N}_i(t)| \to 0$ as $t \to +\infty$ for all i, then $\overline{N}_i$ is called asymptotically stable (which we will abbreviate as A.S.).

Our sole concern in these notes, as far as stability is concerned, will be either with asymptotic stability or with instability (i.e. when $\overline{N}_i$ is not stable in the sense of the negation of the above definition).

If, in order to follow the usual linearization procedure, we define $x_i = N_i - \overline{N}_i$ and substitute $N_i = x_i + \overline{N}_i$ into (1.1), we will then wish to ignore all resulting terms of order two or more in the x_i variables and study the resulting linear system. Before discussing the stability of linear integrodifferential systems we pause for a few words concerning the justification for this linearization procedure for integrodifferential systems.

Suppose we introduce the matrix notation $x = \mathrm{col}(x_i)$, $N = \mathrm{col}(N_i)$ and $\overline{N} = \mathrm{col}(\overline{N}_i)$. (All unsubscripted dependent variables in these notes will denote matrices unless otherwise stated.) Then the change of dependent variables as described above will in general result in a system of the form

$$x' = A(t)x + \int_{-\infty}^{t} B(t,s)x(s)ds + g(t,x)(t) \tag{2.1}$$

where A and B are $n \times n$ matrices which depend on $\overline{N}$ (as well as f_i) and where g is "higher order" in x. Clearly the instability or the A.S. of the zero solution $x \equiv 0$ of (2.1) is equivalent to that of $\overline{N}$ as a solution of (1.1). The question is whether the instability or A.S. of the zero solution $x \equiv 0$ of (2.1) is guaranteed by that of the zero solution $y \equiv 0$ of the linearized system

$$y' = A(t)y + \int_{0}^{t} B(t,s)y(s)ds \tag{2.2}$$

where we have now ignored not only higher order terms in x but a nonhomogeneous (forcing) term involving the initial conditions $x^0 \equiv N_i^0(t) - \overline{N}_i(t)$, $t \varepsilon R^-$; namely we have dropped the expression

$$\int_{-\infty}^{0} B(t,s)x^0(s)ds + g(t,x)(t) \tag{2.3}$$

from (2.1). Conditions which allow an affirmative answer to this question are known (Cushing (1975)). These conditions, which are interpretable in terms of the instability or A.S. of the zero solution of (2.2), are difficult however to relate directly to A and B in any general way. This is not surprising since even for the nondelay case $B \equiv 0$ the A.S. or instability of the nonautonomous ordinary differential system (2.2) is difficult in general to determine from the coefficient matrix A. Of course in this special case of ordinary differential systems the autonomous case when A is a constant matrix is easily handled, at least in principle, since it is reducible to algebraic techniques. This turns out to be true, as we shall see in the next section, when A is a constant matrix and $B(t,s) \equiv B(t-s)$, $\int_0^\infty |B(s)|ds < +\infty$, a case which results when one linearizes an autonomous integrodifferential system about an equilibrium $\overline{N}_i \equiv e_i$.

It should be remembered that whereas A.S. for the zero solution of (2.2) is global in the sense that all solutions tend to zero as $t \to +\infty$, as it always is for linear systems, the A.S. of the zero solution of the perturbation (2.1) is in general only local.

2.2 Autonomous Linear Systems. In studying the stability of equilibria we will encounter systems of the form

$$y' = Ay + \int_0^t B(t-s)y(s)ds \tag{2.4}$$

where A is a constant $n \times n$ matrix and $B(s)$ is a piece-wise continuous $n \times n$ matrix whose entries lie in L^1. For such systems it has been proved (Miller (1972)) that (2.4) is A.S. (i.e. the zero solution is A.S.) if and only if

$$D(z) :\equiv \det(zI - A - B^*(z)) \neq 0 \quad \text{when} \quad \operatorname{Re} z \geq 0. \tag{2.5}$$

Here

$$B^*(z) := \int_0^\infty e^{-zs}B(s)ds$$

the Laplace transform of the kernel B. This condition, involving the so-called racteristic function $D(z)$ of the system (2.4), is of course a generalization the familiar asymptotic stability condition for autonomous differential systems $\equiv 0$) which requires that the eigenvalues of the coefficient matrix A all lie the left half complex plane $\operatorname{Re} z < 0$.

If this stability condition (2.5) is met then the zero solution of the per- bed system (2.1) is (locally) A.S. (Miller (1972), Cushing (1975)). This is ause first of all the $g(x)$ term in (2.3) is higher order in x and secondly ause the first term in (2.3) is small, if the initial function x^0 is small, also tends to zero as $t \to +\infty$. This is shown by the bounds

$$\left| \int_{-\infty}^{0} B(t-s)x^0(s)ds \right| \leq \delta \int_{t}^{+\infty} |B(s)|ds$$

$|x^0(s)| \leq \delta$, $s \leq 0$.

For kernels $B(s)$ constructed from elementary functions $D(z)$ is usually a ional function in z and the stability condition can be tested by means of the l-known Routh-Hurwitz criteria.

If as mentioned in Chapter 1, Stieltjes integrals are used in the model (1.1) h integrators constant after some fixed time T so that the delay integrals are the form

$$\int_{0}^{T} g(N(t-s))dh(s) = \int_{-T}^{0} g(N(t+s))dh(s)$$

n the linearization procedure leads to a linear system of the form

$$y' = \int_{-T}^{0} dC(s)y(t+s).$$

h functional equations have been extensively studied and as is well known such a system is A.S. if

$$(2.6) \qquad \Delta(z) := \det\left(zI - \int_{-T}^{0} e^{zs}\, dC(s)\right) \neq 0 \quad \text{when} \quad \operatorname{Re} z \geq 0$$

holds (Hale (1971)). The main difference in dealing with the two models is that (2.6) is usually a transcendental equation (even for simple integrators $C(s)$) whose roots are accordingly more difficult to locate in the complex plane.

Returning to the Volterra model (2.4) we are also interested in the questior of the instability of the zero solution and to what extent this implies the insta bility of the zero solution of (2.1). If $D(z)$ has at least one root in the rig half complex plane then the linear system (2.4) is unstable (Miller (1974)). The extent to which this instability is present in the perturbed system (2.1) does nc seem to have been completely characterized. The instability of (2.1) has been established under certain conditions on the number of roots in the right half pla and on their multiplicities (Miller (1974), Cushing (1975)) and these results wil suffice in most of our specific models below. It seems reasonable, however, to assert the instability of (2.1) whenever $D(z)$ has at least one root in the righ half plane under no additional constraints and we will in fact do so when such an occasion arises.

CHAPTER 3. STABILITY AND DELAY MODELS FOR A SINGLE SPECIES

Suppose that the functional equation

$$N'/N = f(N)(t)$$

used to describe the growth of a single species whose density is $N = N(t)$. e functional expression f is assumed (as always) to be defined at least r functions $N \geq 0$. If f is independent of N then the model says that the r unit growth rate of the species is unaffected by the density of the species hich is perhaps reasonable for small densities or at least when resources are rge, but certainly is not reasonable in a resource limited environment). For is reason we defined $f(0)(t)$ to be the natural or inherent per unit growth te (net birth rate or death rate if it is positive or negative respectively) nce it would be the growth rate obtained from the model if the dependence on were ignored (i.e. if unlimited resources were available). Since it is usually sumed that a species' own population size can only have a negative effect on its owth rate, the functional f is usually assumed to satisfy $df/dN \leq 0$ at least r large N (with a proper interpretation of df/dN for functional expressions, y as a Fréchet derivative).

Certainly the most famous example is the familiar logistic equation, a non- lay model obtained by choosing f to be a linear expression in N:

$$N'/N = b - aN, \quad b > 0, \quad a > 0.$$

is differential equation is easily integrated to obtain

$$N(t) = \frac{N(0)\exp(\int_0^t b\,ds)}{1 + N(0)\int_0^t a\exp(\int_0^r b\,ds)dr} .$$

If b and a are constants then one finds that all solutions with $N(0) > 0$ tend to $e = b/a$ as $t \to +\infty$ and consequently e is called the carrying capacity or saturation value of the species' environment. For a discussion of the underlying biological assumptions inherent in this model see Pielou (1969, p. 30).

In this chapter we will investigate the stability or instability of not only the logistic model but also more general models under the assumption that time delays are present in the response of the growth rate to changes in species density. (As an illustration of how time delays can significantly affect the growth of populations see the discussion in F. Smith (1963).) One predominant theme will be that time delays tend not to change the stability or instability properties of an equilibrium unless they are in some sense "significant." The "significance" of the time delay may be defined in many ways. Here we will consider sometimes the "magnitude" of the effect on growth rate response caused by the delay, at other times the "length" of the delay or even at times the manner in which the effects are distributed into the past (e.g. monotonically decreasing, etc.).

3.1 Delay Logistic Equations. Consider the integrodifferential equation

$$N'/N = b - aN - d\int_{-\infty}^{t} N(s)k(t-s)ds$$

(3.1)

$$b > 0; \quad a \text{ and } d \geq 0; \quad a + d \neq 0; \quad k \in L_{+}^{1}; \quad |k|_1 = 1.$$

Here we have separated the dependence of f on N into two parts, a nondelay term $-aN$ and a delay term represented by the Volterra integral. We do this in order to discuss their relative importance.

We suppose in this section that b, a and d are all constants. Then (3.1) has a unique positive equilibrium given by

$$e = b/(a + d) > 0.$$

this equilibrium $N \equiv e$ is A.S. then it would again be natural to call it the rrying capacity of the environment.

Letting $x = N - e$ and following the linearization procedure discussed in apter 2 we arrive at the linear equation

$$x' = -eax - ed \int_0^t x(s)k(t - s)ds \tag{.2}$$

ose characteristic equation is

$$D(z): = z + ea + edk^*(z) = 0$$

$$k^*(z): = \int_0^\infty e^{-zt}k(t)dt.$$

we assume

$$a > d \tag{.3}$$

n since $|k^*(z)| \leq 1$ for $\text{Re } z \geq 0$ we find that

$$|z + ea| \geq ea > ed \geq |edk^*(z)|, \quad \text{Re } z \geq 0$$

hence $D(z)$ cannot vanish for $\text{Re } z \geq 0$. This proves the following result.

THEOREM 3.1 *The equilibrium* $e = b/(a + d)$ *is (locally) A.S. as a solution* (3.1) *provided* (3.3) *holds*.

Note that (3.3) says that the delay effect is not too large in the sense that its "magnitude" d is less than that a of the instantaneous effects.

Although this linearization approach yields only the local A.S. of (3.1), the A.S. for this model is in fact global. It was proved by Miller (1966) that any positive solution of (3.1) tends to the equilibrium e as $t \to +\infty$ when (3.3) holds.

Equation (3.1) however may still possess a stable equilibrium even when (3.3) does not hold, that is even if the instantaneous effects on the growth rate are less in magnitude than those subject to delays. This can occur for example if the length of time needed for the maximum response to a change in density to be felt is not too large. To see this consider the model (3.1) with $a = 0 < d$ (violating (3.3)) and with the generic delay kernel $k(t) \equiv T^{-2}t\exp(-t/T)$, $T > 0$. In this case we have a model which assumes negligible instantaneous response and assumes that the delay effects are of "magnitude" d with the maximum influence at any time t being due to density at time $t - T$. The equilibrium is $e = b/d$ and the characteristic equation of the linearized model (3.2) is

$$D(z) := \frac{T^2z^3 + 2Tz^2 + z + b}{(Tz + 1)^2} = 0.$$

A straightforward application of the Hurwitz criteria to the numerator of $D(z)$ shows that there are no roots lying in the right half plane if and only if $T < 2/b$. Thus <u>the equilibrium</u> $e = b/d$ <u>is (locally) A.S. for the generic delay kernel</u> $k(t) \equiv T^{-2}t\exp(-t/T)$ <u>with relatively "short" delays</u> $T < 2/b$ <u>in</u> (3.1) <u>with</u> $a = 0$ <u>and is unstable for "long" delays</u> $T > 2/b$. This result is a special case of the same result for more general models as will be discussed in Section 3.2 below.

If, on the other hand, the delay in the growth rate response is "too large" either in terms of the "magnitude" of the delayed effects or in terms of the "time th of the delay" then one expects (and gets) oscillations from the model). This phenomenon will be considered later in Chapter 5. This is in gen- accompanied by a destabilization of the equilibrium as can be seen in the ple above where e becomes unstable for $T > 2/b$.

Another way of viewing the above simple result is to recognize two time es: $1/b$, that of the inherent rate at which the unrestricted population s, and T, that of the length of delay in the growth rate response to density ges. The stability then depends on the relative sizes of these time scales as ured by the dimensionless parameter bT. This is the point of view of many ussions of time delays in growth models found in the literature (see for exam- May et al. (1974) and J. M. Smith (1968)) and is the point of view we will n take in these notes.

Whether e is stable or not it turns out that we can say some general things t positive solutions of (3.1).

(i) <u>Any positive solution of (3.1) with $a > 0$ is bounded in the future</u>; is, if $N > 0$ for all $t\varepsilon R$ satisfies (3.1) then there exists a constant 0 such that $0 < N(t) \leq M$ for all $t\varepsilon R^+$. To see this observe that

$$N'/N \leq b - aN \quad \text{for} \quad t\varepsilon R^+ \tag{}$$

suppose that N were unbounded. If $N \to +\infty$ as $t \to +\infty$, then (3.4) implies absurdity that $N' < 0$ for all large t. Thus, if N is to be unbounded it oscillate in the sense that $0 \leq \liminf N < \limsup N = +\infty$. But then there ts a $t'\varepsilon R^+$ for which $N'(t') = 0$ and $N(t') > b/a$ which, together with), implies the absurdity $0 < 0$.

(ii) Secondly we can say something about the long term average

$$A(N) := \lim_{t \to +\infty} t^{-1}\int_0^t N(t)dt$$

of any positive solution of (3.1) which is bounded away from zero. Only such solutions are of interest since solutions close to zero violate the assumption that the population size is large which is made when using models based differential equations. It is not clear mathematically however when positive solutions of (3.1) are bounded away from zero, especially when significant delays are present and as a result solutions have large oscillations.

Suppose N is a bounded, positive solution of (3.1) for which $N(t) \geq \delta > 0$ for all $t \in R^+$ and some constant δ. Then

$$0 < \delta \leq N(t) \leq M \quad \text{for} \quad t \in R. \tag{3.5}$$

Let $\varepsilon > 0$ be an arbitrary but fixed real and choose $t' > 0$ so large that

$$1 - \varepsilon < \int_0^t k(s)ds \leq 1 \quad \text{for} \quad t \geq t'. \tag{3.6}$$

We assume that the initial function $N^0(t) = N(t)$, $t \in R^-$ is bounded: $0 \leq N^0(t) \leq M$, $t \in R^-$.

Integrating (3.1) from $t = 0$ to $t = t^* > t'$ we obtain

$$a \int_0^{t^*} N(s)ds = bt^* - dI_1 - \ln N(t^*)/N(0) \tag{3.7}$$

where

$$I_1 := I_2 + \int_0^{t^*} \int_{-\infty}^{-t'} k(t - s)N(s)dsdt$$

$$I_2 := \int_0^{t^*} \int_{-t'}^{t} k(t - s)N(s)dsdt.$$

A straightforward calculation yields

$$I_2 = \int_{-t'}^{0} \int_{0}^{t*} k(t - s)dtN(s)ds + \int_{0}^{t*-t'} \int_{0}^{t*-s} k(w)dwN(s)ds$$

$$+ \int_{t*-t'}^{t*} \int_{0}^{t*-s} k(w)dwN(s)ds.$$

us $k(s) \geq 0$ and $\int_0^\infty k(s)ds = 1$ easily imply

$$I_2 \leq \int_{-t'}^{0} N^0(s)ds + \int_{0}^{t*} N(s)ds \tag{.8}$$

d, since in the middle integral for I_2 we have $t* - s \geq t'$, it follows from .6) that

$$I_2 \geq (1 - \varepsilon) \int_{0}^{t*-t'} N(s)ds = (1 - \varepsilon) \int_{0}^{t*} N(s)ds - (1 - \varepsilon) \int_{t*-t'}^{t*} N(s)ds.$$

om this estimate and (3.7) we obtain

$$(a + d(1 - \varepsilon)) \int_{0}^{t*} N(s)ds \leq bt* + d(1 - \varepsilon) \int_{t*-t'}^{t*} N(s)ds - \ln N(t*)/N(0).$$

wever (3.5) implies that

$$\frac{1}{t*} \int_{t*-t'}^{t*} N(s)ds \to 0 \quad \text{and} \quad \ln N(t*)/N(0) \to 0$$

$t* \to +\infty$ and hence

$$\limsup_{t* \to +\infty} \frac{1}{t*} \int_{0}^{t*} N(s)ds \leq \frac{b}{a + d(1 - \varepsilon)} . \tag{.9}$$

rthermore (3.8) and (3.7) also imply that

$$(a + d)\int_0^{t^*} N(s)ds \geq bt^* - d\int_0^{t^*}\int_{-\infty}^{-t'} k(t - s)N(s)dsdt - d\int_{-t'}^{0} N(s)ds - \ln N(t^*)/N(0).$$

By (3.6)

$$\int_0^{t^*}\int_{-\infty}^{-t'} k(t - s)N(s)dsdt \leq Mt^* \int_{t+t'}^{\infty} k(s)ds \leq M\varepsilon t^*$$

and thus

$$(a + d) \int_0^{t^*} N(s)ds \geq bt^* - M\varepsilon t^* - d \int_{-t'}^{0} N(s)ds - \ln N(t^*)/N(0)$$

which implies

(3.10) $$\liminf_{t^* \to +\infty} \frac{1}{t^*} \int_0^{t^*} N(s)ds \geq \frac{b - dM\varepsilon}{a + d} .$$

Since $\varepsilon > 0$ was arbitrary we obtain from (3.9) and (3.10) that both lim sup and lim inf equal $b/(a + d)$ and hence the limit exists and also equals this ratio.

We conclude _that any positive solution of_ (3.1) _which is bounded above and away from zero for_ $t\varepsilon R^+$ _and which has bounded initial values, has a long time average equal to the equilibrium_: $A(N) = b/(a + d)$.

(iii) Finally, _if_ $k(t)$ _has compact support_: $k(t) = 0$, $t < t_0 < 0$ _and_ $N(t)$ _is a solution of_ (3.1), _bounded for_ $t\varepsilon R^+$, _for which_ $N^0(t)$, $t\varepsilon R^-$ _is bounded away from_ b/d: $0 \leq N^0(t) \leq K < b/d$, $t\varepsilon R^-$, _then_ $N(t)$ _is bounded away from zero for all_ $t\varepsilon R^+$.

To prove this statement we observe from (3.1) that

$$N'/N \geq - aM - dK - dM: = \alpha, \quad t\varepsilon R^+$$

which yields by integration the bound

$$N(s) \leq N(t) \exp(-\alpha(t - s)), \qquad t \geq s \geq 0.$$

pose that N is not bounded away from zero and hence that $\liminf N(t) = 0$.

First suppose $t_m > 0$ is a point at which $N(t)$ attains a (local) minimum such exists). Then

$$0 = b - aN(t_m) - d \int_{-\infty}^{t_m} k(t_m - s)N(s)ds \geq b - aN(t_m) - dK - dN(t_m)\beta$$

$$\beta: = \int_0^{t_0} k(s) \exp(-\alpha s)ds > 0$$

ch in turn yields

11) $$N(t_m) \geq (b - dK)/(a + d\beta) > 0.$$

ce the right hand side of (3.11) is a constant we see that $N(t)$ cannot have sequence of minima $t_m \to +\infty$ such that $N(t_m) \to 0$. This fact together with the sumption that $\liminf N(t) = 0$ implies that $N(t) \to 0$ as $t \to +\infty$. As a re- t we can find a $t^* > 0$ so large that $0 \leq N(t) \leq b/2(a + d)$ for $t \geq t^*$. then from (3.1) for $t \geq t^* + t_0$

$$N'/N \geq b - ab/2(a + d) - db/2(a + d) = b/2 > 0$$

ch implies the contradiction that $N(t) \geq N(t^* + t_0) \exp(bt/2)$, $t \geq t^* + t_0$ d hence $N(t) \to +\infty$ as $t \to +\infty$.

As a result of this contradiction it must be the case that $N(t)$ is bounded ay from zero. □

All three (i)-(iii) of the above properties of positive solutions can be established in the same manner for (3.1) with $k(s)ds$ replaced by $dh(s) \geq 0$.

3.2 The Logistic Equation with a Constant Time Lag. One of the few delay models which has been extensively studied in the mathematical literature is the delay logistic model (3.1) with $a = 0$ and the Volterra integral replaced by the time lag expression $N(t - 1)$:

$$N'/N = b - dN(t - 1), \quad b > 0, \quad d > 0. \tag{3.12}$$

If we let $x = (dN - b)/b$ then (3.12) reduces to

$$x' = -bx(t - 1)(1 + x) \tag{3.13}$$

which is the form in which (3.12) has been most extensively studied. Here the time variable t has been assumed scaled (without loss in generality) so that the time lag is of unit length.

Intuitively the time lag model (3.12) is less reasonable than the (continuously distributed) delay models in Section 3.1 (see Caswell (1972)). There also is a slight amount of experimental evidence supporting this statement (Caperon (1969)). Nonetheless (3.12) has been extensively studied in the literature and it is probably desirable at least very briefly to consider its stability. It will be clearly seen how the more realistic models of Section 3.1 are in fact easier to study since (3.12) has a transcendental characteristic equation.

The linearization of (3.13) about $x \equiv 0$ yields $x' = -bx(t - 1)$ which has characteristic equation (cf. (2.6))

$$z + be^{-z} = 0. \tag{3.14}$$

$z = r + is$. Then equating both real and imaginary parts of $z + be^{-z}$ to
ɔ we find

$$\text{(a)} \qquad r + be^{-r}\cos s = 0$$

15)

$$\text{(b)} \qquad s - be^{-r}\sin s = 0.$$

Consider first the possibility of a real root $z = r$ for which (3.15b) holds ɔmatically. It is clear graphically that $r + be^{-r} = 0$ has either no real ts or two negative roots the two cases being separated by the value of b for ɔh there is a double root. Since a double root can only occur when $1 = be^{-r}$ find that the only possible double root is $r = -1$ which occurs for $b = 1/e$. s for $b < 1/e$ we find that (3.14) has two negative real roots while if $b > 1/e$ ɩas no real roots.

Suppose we look for roots $z = r + is$, $s \neq 0$. Eliminating be^{-r} from (3.15) get

16) $$r = -s\cot s$$

ɔh when substituted back into (3.15b) yields

17) $$s/b = e^{s\cot s}\sin s.$$

ɔlex roots of (3.14) are found by solving (3.17) for s and obtaining r from ɔy (3.16). Since roots of (3.14) appear in complex conjugate pairs we only sider $s > 0$. Equation (3.17) for $s > 0$ may be studied graphically by inter- ɔing the curve $y = e^{s\cot s}\sin s$ drawn in FIGURE 3.1 with the straight line s/b.

If $b < 1/e$ we obtain no root s of (3.9) in $(0,\pi)$, but a root s in each interval $(2\pi, 5\pi/2)$, $(4\pi, 9\pi/3)$, etc. Each of these yields from (3.16) a corresponding $r < 0$. Thus if $b > 1/e$ we have two real negative roots and infinitely many complex roots of the characteristic equation (3.14) lying in the left half plane only.

If $1/e < b < \pi/2$ the case is as above except that there are no real roots and there is an additional complex root with $0 < s < \pi/2$ and hence $r < 0$.

Finally if $b > \pi/2$ then there is a complex root with $\pi/2 < s < \pi$ and henc $r > 0$. (A bound for r may be found in Hadeler (1976).)

Consequently when $b < \pi/2$ the original model (3.12) has a (locally) A.S. equilibrium $e = b/d$. When $b > \pi/2$ this equilibrium is unstable.

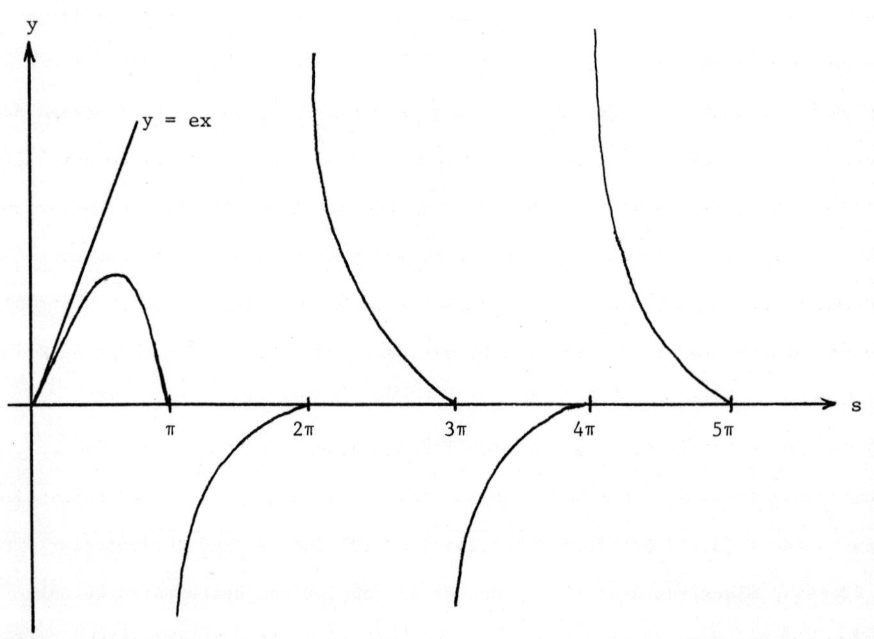

FIGURE 3.1

The model (3.12) has been studied in much greater detail than we have done re. Wright (1955) proved the global A.S. of the equilibrium for $b < 3/2$. kutani and Markus (1958) showed that all solutions oscillate for $b > 1/e$ and not oscillate for $b < 1/e$. Many studies have also been made of (3.12) for $> \pi/2$ (and more general models with a time lag) concerning the existence of riodic solutions (see for example Jones (1962a, 1962b), Kaplan and Yorke (1975), d Nussbaum (1973)).

3.3 Some Other Models. It is clear that any model $N'/N = f(N)(t)$ involving lays of the Volterra integral type and possessing an equilibrium can be investi- ted locally near $N = e$ using the same techniques of linearization as above. e model may even possess more than one equilibrium, each of which is treated sep- ately.

The model (3.1) investigated in Section 3.1 above is one in which f (i.e. e growth rate response) is a linear function of density N. More detailed models ich attempt to take into account the nature of the inhibition due to population nsity and possible delays may involve higher order terms in f. As one example this we consider a model in which the delayed growth rate responses are related accumulating environmental intoxicants due, say, to catabolic waste residuals. ese pollutants are in turn related to past population sizes. A model of this pe was proposed and numerically studied by Borsellino and Torre (1974). To sim- ify the ideas (while not, it turns out, ruining some of the qualitative features their model) we will consider a model which is similar in form and motivation to at of Borsellino and Torre, but which is in its details considerably simpler and re manageable analytically.

One of the features of the model of Borsellino and Torre is that the coeffi- ent d in (3.1), which measures the magnitude of the response of the growth rate past population sizes, is a function of some measure of the accumulated pollution which in turn depends on past population sizes: $P(t) = P_0\int_0^\infty N(t-s)k(s)ds$, $P_0 > 0$. Thus we assume $d = d(P)$. Suppose that we in fact assume that d is proportional to P (Borsellino and Torre assume d is proportional to P^2) so that our model becomes

$$N'/N = b - aN - d_0p_0\left[\int_0^\infty N(t-s)k(s)ds\right]^2. \tag{3.18}$$

Here the term $-aN$ is interpreted as the density inhibition term and accounts for the deleterious effects of crowding (whatever they might be) while the integra terms account for the effects of the accumulated pollutants. The growth rate response to crowding is assumed to have negligible delay compared to the delay in response to the pollutants, an assumption which seems particularly appropriate on perhaps a long time scale.

This model has a unique equilibrium $e > 0$ given by the positive root of the quadratic equation $d_0p_0e^2 + ae - b = 0$ and hence

$$e = \frac{-a + (a^2 + 4d_0p_0b)^{1/2}}{2d_0p_0} > 0. \tag{3.19}$$

If we linearize (3.18) about this equilibrium we obtain

$$x' = -aex - 2d_0p_0e\int_0^t x(s)k(t-s)ds \tag{3.20}$$

which has the same form as (3.2) with d replaced by $2d_0p_0$. Thus as in Section 3.1 the equilibrium (3.19) is locally A.S. provided $2d_0p_0 < a$, that is provided the accumulated effects of pollution are not too large. Also in this case $P(t) \to p_0e$ as $t \to +\infty$.

Suppose $k(s) = T^{-2}s\exp(-s/T)$, $T > 0$. (This kernel is a "smoothed out" version of that used by Borsellino and Torre.) Then the characteristic equation

(3.20) is $D(z) = N(z)/(Tz + 1)^2$ where

$$N(z) = T^2z^3 + (T^2ae + 2T)z^2 + (1 + 2Tae)z + (ae + 2d_0p_0e).$$

a = 0 we have just as in Section 3.1 above that the equilibrium is A.S. if $1/d_0p_0e = (d_0p_0b)^{-1/2}$ and unstable if $T > (d_0p_0b)^{-1/2}$. If $a > 0$ then by Hurwitz criteria $N(z)$ has all of its roots in the left half plane if and y if $H(T;a) := a^2e^2T^2 + (2ae - d_0p_0e)T + 1 > 0$. It is straightforward to show t $H > 0$ for all T if $d_0p_0 < 4a$ while if $d_0p_0 > 4a$ then $H < 0$ if and y if $T_1 < T < T_2$ where T_i are the two positive real roots of the quadratic ation $H = 0$. Thus if the pollution coefficient p_0 is large, this model has unusual property that the equilibrium, which is stable for small delays T_1 and unstable for $T_1 < T < T_2$, regains stability for large enough delays T_2.

Note that the equilibrium population e in (3.19) decreases while the pollu-n equilibrium $P_0 = p_0e$ increases with increasing pollution coefficient p_0. o it turns out that the critical delay value T_1 is a monotonically decreasing zero) function of p_0 and as a result the onset of instability occurs for rter delays as p_0 increases. These qualitative features (and others) were nd numerically by Borsellino and Torre for their model.

Many other models have been proposed for describing the growth of a single ulation and any of these could be modified to include delays. For example the l-known equation of Gompertz (1825)

$$N'/N = b \ln (e/N)$$

ld be modified in a manner analogous to that used for the logistic equation to lude delays in the density term

$$N'/N = b \ln e - b \int_0^\infty k(s) \ln N(t - s)ds.$$

Clearly this model is linear in $\ln N$ and as such is amenable to stability analysis straight away (the results in this case being global). Models of the form

$$N'/N = b - dN^m, \quad m > 0$$

have also been proposed and used (for example see Gilpin and Ayala (1973)). These could be modified to include delay terms such as $(\int_0^\infty N(t - s)k(s)ds)^m$ or $\int_0^\infty N^m(t - s)k(s)ds$ in place of N^m. The local stability analysis around the equilibrium $e = (b/d)^{1/m}$ of the resulting models would again yield the linear equation (3.2) with $(a = 0)$ whose local stability analysis would then be no different from that of (3.1).

3.4 Some General Results. Consider the general model

(3.21) $$N'/N = bf(N)(t)$$

where b = constant > 0 and f is a functional defined for nonnegative functions $N \geq 0$ and satisfies $f(e) = 0$ for some constant $e > 0$. Here we think of b as the inherent net per unit growth rate which the population would have in the absence of density restraints (i.e. when $f \equiv 1$). Clearly e is an equilibrium. We also assume that f has enough smoothness at $N \equiv e$ so that (3.21) may be linearized at $N \equiv e$ to yield a linear equation of the form

$$x' + ab \int_0^t x(s)k(t - s)ds = 0, \quad k \varepsilon L_+^1, \quad |k|_1 = 1$$

for some constant $a > 0$. Since a is essentially e times $-f'(e)$ this assump on requires that $f'(e) < 0$, which means that in the neighborhood of the uilibrium e increased density results in a decreased growth rate. These re-rictions are fulfilled for example by the models considered in the Sections 1 and 3.3 above.

We wish first to study the stability of the equilibrium e as it is depen-nt on the relationship between the time scale $1/b$ of the inherent growth rate d a time scale determined by the delay in the growth rate response to density anges. As we saw in Section 3.1 (and as is often done in the literature, e.g. May et al. (1974) and J. M. Smith (1968)) the dimensionless parameter bT ere T is some measure of the response delay can often conveniently be used r this purpose. In the general setting of model (3.21) let us then let $T > 0$ some measure of the response delay described by the functional, or more spe-fically by the delay kernel $k(t)$. For example, if $k(t)$ has a unique maximum might be chosen to be that value of t at which k assumes this maximum (as r example in the generic kernel $k(t) = T^{-2}t \exp(-t/T)$); or T might be osen as $T = \int_0^\infty tk(t)dt$, the first moment of k, or such that $\int_0^T k(t)dt = 1/2$. thematically it doesn't matter, at least for our purposes, how T is chosen. simply assume that it can be defined in some meaningful way once $k(t)$ is ven.

We wish to show for this general model that the equilibrium e is (locally) S. at least for bT sufficiently small. In the following Section 3.5 we con-der the question of instability for bT large.

If, in order to introduce the delay measure T explicitly into the analysis, new time scale $t^* = t/T$ is chosen and the model (3.21) is transformed into one r the unknown function $N^*(t^*) := N(t^*T)$, the result is an equation of the same rm with b replaced by bT and with a new delay kernel $k^*(t^*) \equiv Tk(t^*T)$ ich has delay measure $T^* = 1$ and still satisfies $k^* \varepsilon L_+^1$, $|k^*|_1 = 1$.

Thus, without loss in generality we consider the model

$$N'/N = bTf(N)(t) \tag{3.22}$$

where in addition to the hypotheses on f and k made above we have that the delay measure is 1. The linearization about e then becomes

$$x' + abT\int_0^t k(t - s)x(s)ds = 0. \tag{3.23}$$

The characteristic equation of this linear equation is

$$D(z,u) := z + uk^*(z) = 0, \quad u = abT > 0. \tag{3.24}$$

THEOREM 3.2 *If, in addition to the assumptions on* f *and* k *made above, it is assumed that* $tk(t)\varepsilon L^1$ *then the equilibrium* e *is locally A.S. as a solution of the general model* (3.21) *provided* bT *is sufficiently small.*

Proof. We wish to show that for $u > 0$ small D has no roots with $\mathrm{Re}\, z \geq 0$. Suppose this were not true. Then we could find sequences z_n, u_n such that $\mathrm{Re}\, z_n \geq 0$, $u_n > 0$, $u_n \to 0$ and $D(z_n,u_n) = 0$. Then

$$|z_n| \leq |u_n\int_0^\infty e^{-z_n t}k(t)dt| \leq u_n$$

shows that $z_n \to 0$ as $n \to +\infty$.

However, if we apply the implicit function theorem to the equation (3.24), observing that $D(0,0) = 0$, $D_z(0,0) = 1$ (here we use $tk(t)\varepsilon L^1$), we obtain a unique solution branch $z = z(u)$ for small u. An implicit differentiation of $D(z(u),u) = 0$ yields $z'(0) = -1$ so that $\mathrm{Re}\, z(u) < 0$ for $u > 0$ small.

The sequences z_n, u_n contradict the uniqueness of the branch $z(u)$ obtained from the implicit function theorem. □

Another point of view which might be taken with respect to the delay logis-c (3.21), as opposed to that above of comparing the time scale of the inherent rth rate to that of the delay in growth rate response, is that of studying the ability of its equilibrium as it depends on the weighted manner in which the owth rate responds to past population densities; that is to say, the "shape" the delay kernel $k(t)$. For example, if $k(t) \varepsilon L^1_+$ is monotonically decreasing e might expect asymptotic stability since the nondelay logistic has an asymp-tically stable equilibrium and since, for such a delay kernel, the maximum owth rate response to density changes would be instantaneous while the delayed sponse to past population densities would be monotonically decreasing. This s in fact been shown to be true by Walther (1976) provided $k(t)$ is also con-x with compact support.

Our approach in the next theorem is by way of the Argument Principle applied the characteristic function of the linearization (6.21). Since this approach n be generalized and applied to systems, we postpone the details until later. e following theorem is then a corollary of Theorem 4.12 which is proved in apter 4.

THEOREM 3.3 *Suppose that the above assumptions on* f *hold in* (3.21) *and* *at* $k(t) \varepsilon L^1_+$, $tk(t) \varepsilon L^1$, $|k|_1 = 1$. *Suppose that the characteristic function* $z): = z + abk^*(z)$ *of* (3.21) *has no purely imaginary roots. Then* $\lim_{R \to +\infty} D(iR)$ *uals* $(1 + 4k)\pi/2$ *for some integer* $k = 0, -1, -2, \ldots$ *and the equilibrium of* .21) *is*

(a) *(locally) A.S. if* $k = 0$

(b) *unstable if* $k < 0$.

If we denote

$$C(R): = \int_0^\infty k(t) \cos Rtdt, \qquad S(R): = \int_0^\infty k(t) \sin Rtdt$$

then $D(iR) = abC(R) + i(R - abS(R))$ and clearly $D(0) = ab > 0$, $|\text{Re } D(iR)| \leq ab$. Thus, we see geometrically (cf. FIGURE 3.2) that the equilibrium is A.S. if and only if the graph of $D(iR)$, $R > 0$ does not "wind around" the origin of the complex plane. Note that $\text{Im } D(iR) \to +\infty$ as $R \to +\infty$ since $S(R)$ is bounded.

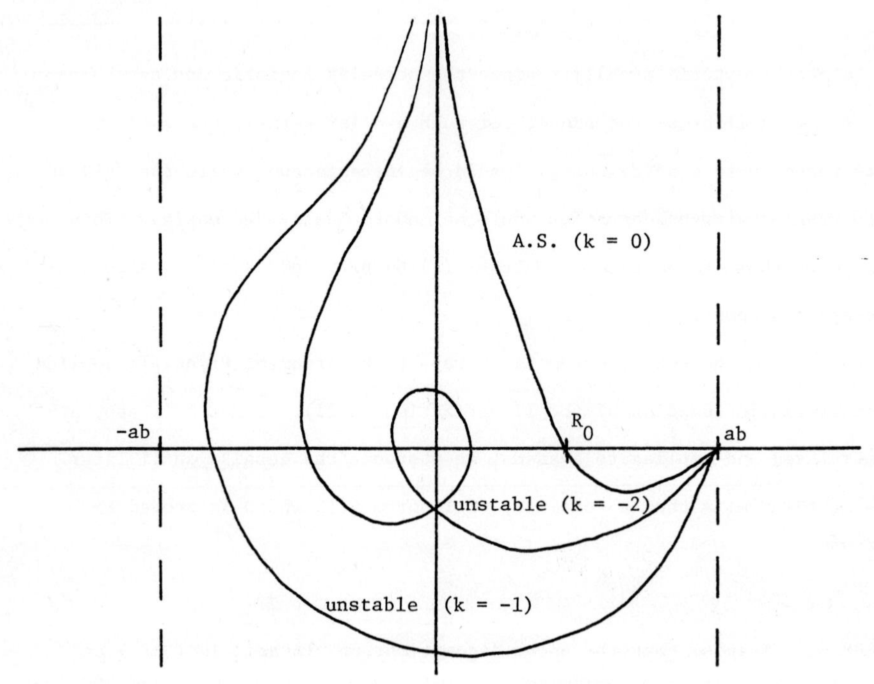

Graph of $D(iR)$, $R > 0$ in the complex plane.

FIGURE 3.2

The following corollary is obvious from the geometric interpretation of Theorem 3.3 (see FIGURE 3.2).

COROLLARY 3.4 _Under the assumptions of Theorem_ 3.3 _the equilibrium of_ (3.21) _is (locally) A.S. if any one of the following conditions hold:_

(a) $C(R) \geq 0$ for all $R > 0$

(b) _if for all roots_ $R = R_0 > 0$ _of_ $C(R) = 0$ _it is true that_

$R_0 > abS(R_0)$

(c) $C(R) \neq 0$ for $0 \leq R \leq ab$.

Part (c) follows from (b) since $|S(R)| \leq 1$. From (a) follows the next
rollary which is itself a generalization of Walther's theorem.

COROLLARY 3.5 _Under the assumptions of Theorem_ 3.3 _the equilibrium of_ (3.21)
(locally) A.S. if $k(+\infty) = k'(+\infty) = 0$ _and_ $k''(t) \geq 0$ _for_ $t \geq 0$.

Proof. Note that the assumptions on $k(t)$ imply that $k''(t)\varepsilon L^1_+$ and
(t) < 0. Two integrations by parts yield, for $R > 0$

$$C(R) = R^{-2} \int_0^\infty k''(t)(1 - \cos Rt)dt > 0. \quad \square$$

This corollary applies for example to the "weak" generic kernel
:) $= T^{-1} \exp(-t/T)$, $T > 0$.

3.5 _A General Instability Result_. Consider again the general model (3.21)
the form (3.22). After linearizing about the equilibrium we again turn our
tention to the characteristic equation (3.24). We saw above that if $u = abT$
small then (3.24) has no roots $\text{Re } z \geq 0$. As u increases we might expect to
counter a value $u = u_0 > 0$ at which $D(z,u_0)$ has a root on the imaginary
is $z = iR$. (That this does not always happen can be seen by the case of the
eric "weak" kernel $k(t) = T^{-1}e^{-t/T}$, $T > 0$.) Note that $\overline{D(z,u_0)} = D(\bar{z},u_0)$
that $z = -iy$ is also a root. We wish to ask in this section under what con-
tions will $D(z,u_0)$ have roots z with $\text{Re } z > 0$ for u near u_0. Under
ch conditions the linear equation (3.23) will be unstable.

First we observe that $D(z,u) = 0$ has a purely imaginary root $z = iR$ for some $u = u_0 > 0$ if and only if

$$(3.25)\quad C(R) := \int_0^\infty k(t)\cos Rt\,dt = 0, \quad S(R) := \int_0^\infty k(t)\sin Rt\,dt > 0 \quad \text{for some } R > 0$$

in which case

$$(3.26)\qquad u_0 = R/S(R).$$

To find roots z of $D(z,u) = 0$ near $z = iR$, $u = u_0$ we invoke the implicit function theorem. If $tk(t)\varepsilon L^1$ and

$$(3.27)\qquad S^1(R) := \int_0^\infty tk(t)\sin Rt\,dt \neq 0$$

then

$$0 \neq D_z(iR,u_0) = \left[1 - u_0\int_0^\infty tk(t)\cos Rt\,dt\right] + i\left[u_0\int_0^\infty tk(t)\sin Rt\,dt\right]$$

and hence we can solve the characteristic equation for $z = z(u)$ at least for u near u_0 where $z(u_0) = iR$. By implicit differentiation $z'(u_0) = -D_u(iR,u_0)/D_z(iR,u_0)$ which yields

$$\operatorname{Re} z'(u_0) = u_0 S^1(R)/|D_z(iR,u_0)|^2.$$

As a result $\operatorname{Re} z(u) > 0$ for $u > u_0$ (or $< u_0$) near u_0 when $S^1(R) > 0$ (or < 0).

THEOREM 3.6 <u>Suppose that the delay kernel</u> $k(t)$ <u>in the linearization</u> (3.23) <u>of the general single species model</u> (3.22) <u>satisfies</u> $k(t)\varepsilon L^1_+$, $tk(t)\varepsilon L^1$,

$|k|_1 = 1$ *and* (3.25), (3.27) *for some* $R > 0$. *Then the equilibrium* $N \equiv e$ *of* (3.21) *is unstable for* abT *near, but* $abT > u_0$ (*or* $< u_0$) *where* u_0 *is given by* (3.26) *depending on whether* $S^1(R) > 0$ (*or* < 0) *respectively*.

As examples, consider the two generic delay kernels. If $k(t) = \exp(-t)$ then $C(R) = 1/(1 + R^2) \neq 0$ for all R and hence (3.25) cannot be fulfilled. This is of course commensurate with Corollary 3.5. If on the other hand $k(t) = t \exp(-t)$ then

$$C(R) = \frac{1 - R^2}{(1 + R^2)^2}, \qquad S(R) = \frac{2R}{(1 + R^2)^2}$$

so that (3.25) is fulfilled for (and only for) $R = 1$. Since $S^1(1) = 1/2$ Theorem 3.6 applies with $u_0 = 1/S(1) = 2$ and the equilibrium of the general model (3.21) is unstable for abT near but greater than 2. For the special case when the response function f is linear and (3.21) reduces to the delay logistic (for which $e = 1$) this result is completely in accord with the result found in Section 3.1. Theorem 3.6 is generalized in Chapter 4 (cf. Section 4.9, Corollary 4.16).

3.6 The Stabilizing Effect of Delays. The general point of view taken so far in this chapter has been that time delays in a species growth rate response to its own density changes tend to have a destabilizing effect. Thus, an equilibrium which is stable in the absence of such delays is unstable in the presence of delays which are in some sense "significant" (e.g., delays for which the "length" of delay is large compared to the inherent growth rate of the species). In this section we briefly consider what stabilizing effects delays in growth rate response could possibly have.

Consider the general model (3.21), which we restate here for convenience:

(3.21) $$N'/N = bf(N)(t), \quad b > 0$$

which as above we assume has a positive equilibrium $e > 0$: $f(e)(t) \equiv 0$ about which the linearization of (3.21) takes the form

(3.28) $$x' + ab\int_0^t x(s)k(t-s)ds = 0, \quad k \in L^1_+, \quad |k|_1 = 1$$

for some constant $a \neq 0$. Here a is essentially $-ef'(e)$. We do not, however, (as we did above) assume necessarily that $a > 0$. The characteristic function of (3.28) is $D(z) := z + abk^*(z)$.

If there is no delay in this model or more specifically if there is no delay in the linearization (3.28) (formally, $k(s) = \delta_0(s)$), then $D(z) = z + ab$ and the equilibrium is A.S. if $a > 0$ and unstable if $a < 0$. In all of the work in previous sections it was assumed that $a > 0$ and hence that the nondelay version of the model had a stable equilibrium. We then studied to what extent the presence of delays caused the equilibrium to become unstable. As generic cases we saw that if $k(t)$ is the "strong" generic delay kernel $k(t) = T^{-2}t\exp(-t/T)$, $T > 0$, then the equilibrium became unstable for $T > 2/b$, i.e. for long delays. On the other hand, if $k(t)$ is the "weak" generic kernel $k(t) = T^{-1}\exp(-t/T)$, $T > 0$ then the equilibrium remains A.S. for all $T > 0$. However, even in this "weak" case the delay can still be said to have a destabilizing effect in the sense that the approach to equilibrium is slower for larger T. To see this, we compute $D(z) = N(z)/(zT + 1)$ where $N(z) := Tz^2 + z + ab$ and hence find the roots of $z_\pm(T)$ of $D(z)$, as functions of the delay T, to be

(3.29) $$z_\pm(T) = \frac{1}{2T}(-1 \pm (1 - 4abT)^{1/2}).$$

Thus, for large T (namely, $T > 1/4ab$) we see that $\mathrm{Re}\, z_\pm(T) = -1/2T$ which is n increasing (to zero) function of T. Since the approach to equilibrium is at rate determined by the magnitude of this, the largest, negative real part we ee that this approach is slowed as T is increased.

Suppose we turn now to the case $a < 0$ when the equilibrium for the nondelay odel is unstable. First of all we note that <u>the presence of delays will never ause an otherwise unstable equilibrium to become stable</u>. This is because for eal $z > 0$ we find that $D(z) \to +\infty$ as $z \to +\infty$ (since $k^*(z)$ is bounded) hile $D(0) = ab < 0$. Thus, $D(z)$ has at least one positive, real root and as a esult the equilibrium is unstable. Nonetheless, suppose we study the rate that he solutions grow (away from equilibrium) as measured by the largest real part f z, z a root of $D(z)$. Let us do this for the two generic delay kernels.

Suppose $a < 0$ and $k(t) = T^{-2} t \exp(-t/T)$. Then $D(z) = N(z)/(zT + 1)^2$ here

$$N(z) := T^2 z^3 + 2Tz^2 + z + ab.$$

ince $ab < 0$, $N(z)$ has a positive real root. Also $N(z)$ has no purely imagiary roots since $\mathrm{Re}\, D(iR) = -2T + ab < 0$. Finally, the remaining two roots of (z) (and hence of $D(z)$) lie in the left half plane $\mathrm{Re}\, z < 0$. This can be een as follows: if these remaining roots were in the right half plane then all hree roots of $N(z)$ would lie in the right half plane which would imply that ll three roots of

$$N(-z) = -T^2 z^3 + 2Tz^2 - z + ab$$

ould lie in the left half plane. This in turn implies that $N(-z)$ would satisfy he Hurwitz criteria which is obviously false. Thus, N and hence D has two omplex conjugate roots $\mathrm{Re}\, z < 0$ and one positive real root. Let $z = z(T) > 0$

be the positive real root as a function of the delay $T > 0$. An implicit differentiation of

$$N(z(T)) := T^2 z^3(T) + 2Tz^2(T) + z(T) + ab = 0, \quad T > 0$$

yields

$$z'(T) = -2z^2(Tz + 1)/(3T^2z^2 + 4Tz + 1) < 0, \quad T > 0.$$

Thus, $z(T)$ decreases as T increases which can be interpreted as saying that _the instability of the equilibrium of_ (3.21) _when_ $a < 0$ _is weakened as the delay_ T _in the "strong" generic delayed response of the growth rate is increased._

Secondly, suppose $a < 0$ and $k(t) = T^{-1} \exp(-t/T)$, $T > 0$. Then, as above, $D(z) = N(z)/(zT + 1)$ where

$$N(z) = Tz^2 + z + ab$$

which has two real roots $z_- < 0 < z_+$ given by (3.29). Now $z_+ = z_+(T)$ is monotonically decreasing from $z_+(0) = -ab > 0$ to $z_+(+\infty) = 0$. Thus, _the conclusion in the preceding paragraph is valid for the "weak" generic kernel as well._

The idea that time delays can be considered as "stabilizing" in the sense that they weaken the instability of an unstable equilibrium seems to have been first put forth for population dynamical models by Beddington and May (1975). For a specific single species model with a single instantaneous time lag they reached the same conclusions which we reached above for the general model (3.21) with generic delays.

It should be pointed out that a single model may well have both a stable and unstable equilibrium (as in fact Beddington and May's model does). Thus, for such model time delays in the species' growth rate response to population density hanges can cause a weakening of both the stability of the stable equilibrium and he instability of the unstable equilibrium. This would cause the population denity to spend "more time" near the unstable equilibrium as the delay is increased nd hence may be viewed as a stabilizing influence near the unstable equilibrium. eddington and May (1975) cite experimental data which seem to support this conention.

CHAPTER 4. STABILITY AND MULTI-SPECIES INTERACTIONS WITH DELAYS

The purpose of this chapter is to explore briefly some models for species interactions when time delays are present in at least some of the growth rate responses to interactions with either members of other species or the same species. We will confine ourselves mostly to mathematical investigations of the stability of equilibria for models which involve only quadratic interaction terms (i.e. for models (1.1) with f_i linear in N_j). Many ecological models have been proposed and studied in which these response functions f_i are not linear, both for models with delay and (even more so) without. For example models with Michaelis-Menten type terms and with delays have been studied by Caperon (1969) and MacDonald (1976) and with $\ln N_j$ terms with delays by Gomatam and MacDonald (1976). Caswell (1972) numerically studied a delay model involving complicated rational expressions in N_j. In principle however the linearization techniques we use would of course apply to these and any other more general model as well and in fact the results would be identical (allowing for differences in parameter interpretations) for models with identical linearizations. We will also upon occasion consider models in which the response functions are not necessarily linear.

4.1 Volterra's Predator-Prey Model with Delays.

The famous Lotka-Volterra model for a predator-prey interaction is

$$(4.1) \qquad N_1'/N_1 = b_1 - a_{12}N_2, \quad N_2'/N_2 = -b_2 + a_{21}N_1, \quad b_i > 0, \quad a_{ij} > 0.$$

(We will always assume unless otherwise stated that the coefficients in any model under consideration are nonnegative so that the signs appearing in any given model genuinely reflect the nature of the interaction.) Here the f_i in our basic model (1.1) are linear; $\partial f_i/\partial N_i = 0$ so that there is no "self-inhibition" or "resource-limitation" term (sometimes referred to as a "logistic" term); $f_1(0) = b_1 > 0$, $f_2(0) = -b_2 < 0$ implies that the prey species N_1 grows exponentially in the absence of predators N_2 while predators die exponentially in the absence of prey; and $\partial f_1/\partial N_2 = -a_{12} < 0$, $\partial f_2/\partial N_1 = a_{21} > 0$ so that predators inhibit prey growth and prey enhance predator growth.

In his book Volterra (1931) derived a modified version of this model in which he assumes that, while the effect of predators on prey might well be essentially instantaneous, in many interactions the response of predator to contacts with prey may be delayed (due for example to a gestation period). Volterra's delay model is

$$\text{(4.2)} \quad \begin{array}{ll} \text{(a)} & N_1'/N_1 = b_1 - a_{12}N_2 \\ \text{(b)} & N_2'/N_2 = -b_2 + a_{21}\int_{-\infty}^{t} N_1(s)k_1(t-s)ds \end{array}$$

for $k_1 \varepsilon L_+^1$. From now on we assume, unless otherwise stated, that any delay kernel k appearing in a model satisfies the normalization $|k|_1 = 1$. Mathematically, the kernel k_1 serves to describe the weight of the delay effects s time units before t. Volterra's specific description of k is that $k(t) = g(t)h(t)$ where $g(t)$ is the fraction of the population made up of individuals of age greater than t (it is assumed that the age distribution remains constant in time) and $h(t)$ is a resource utilization function. A detailed derivation of this model can be found in Volterra's book or the papers of Rescigno and Richardson (1973) and Scudo (1971).

The delay model (4.2) has the same equilibrium $e_1 = b_2/a_{21} > 0$, $e_2 = b_1/a_{12} > 0$ as that of the nondelay model (4.1). This equilibrium for (4.1) is neutrally stable but not A.S. in that all solutions are periodic and form closed loop trajectories surrounding the equilibrium in the N_1, N_2 phase plane.

We will show that, to the contrary, the delay model (4.2) is "usually" unstable. Volterra's work (which was carried out for a slightly more general model) dealt with oscillatory behavior of solutions and their long term averages (see Chapter 6); however Volterra did not establish the convergence or divergence of these oscillations.

The lack of delay in the growth rate response of the prey as given in equation (4.2a) should be viewed not so much as the complete lack of any delay in this response but rather as saying such delay is significantly less than that in the predator's growth rate response to prey density changes as described by (4.2b). A slightly more general model would be one (which we will denote by (4.2')) in which the equation

$$\text{(a')} \qquad N_1'/N_1 = b_1 - a_{12} \int_{-\infty}^{t} N_2(s)k_2(t - s)ds$$

$k_2 \varepsilon L_+^1$, replaces (4.2a). This model allows for delays in prey growth rate response to predator density and formally yields (4.2) if $k_2(s) = \delta_0(s)$, the Dirac delta function at $s = 0$. If the linearization procedure of Chapter 2 is carried out about the equilibirum $e_1 = b_2/a_{21}$, $e_2 = b_1/a_{12}$ of (4.2') we obtain the linear system

$$x_1' = -e_1 a_{12} \int_0^t k_2(t - s)x_2(s)ds$$

$$x_2' = e_2 a_{21} \int_0^t k_1(t - s)x_1(s)ds$$

whose characteristic equation is $D(z) := z^2 + k(z) = 0$ where $k(z) = b_1 b_2 k_1^*(z) k_2^*(z)$. We need to investigate the possibility that $D(z)$ has roots satisfying $\operatorname{Re} z \geq 0$. Note that if both $k_i = \delta_0$ so that (4.2') reduces to the nondelay Lotka-Volterra model (4.1) then $D(z) = z^2 + b_1 b_2$ has two purely imagry roots as is consistent with above-mentioned neutral stability of this model.

The following theorem shows that the equilibrium of (4.2') is usually unble. Although this theorem is a special case of the more general Theorem 4.12 en and proved later in Section 4.8, we will give a proof for completeness and plicity.

THEOREM 4.1 *Suppose that the delay kernels in* (4.2') *are such that* t)εL^1_+, $tk_j(t)\varepsilon L^1$, $|k_j|_1 = 1$ *and the characteristic function* $D(z) := z^2 +$ $_2k_1^*(z)k_2^*(z)$ *has no purely imaginary roots. Then* $\arg D(+i\infty) = (1 - 2m)\pi$ *for* e *integer* $m = 0, 1, 2, \ldots$ *and the equilibrium* $e_1 = b_2/a_{21}$, $e_2 = b_1/a_{12}$ *is*

(a) *unstable if* $m \neq 0$ *and*

(b) *(locally) A.S. if* $m = 0$.

Proof. The transforms $k_j^*(z)$ are analytic for $\text{Re } z \geq 0$. Let $\partial(R)$ denote boundary of the half circle $\text{Re } z \geq 0$, $|z| = R$ and let $\partial^1(R)$ be the cirar part: $\text{Re } z > 0$. Then $\partial^2(R) = \{z = iy, \ -R \leq y \leq R\}$ and $\partial(R) = \partial^1(R) +$ R). By the Argument Principle the number of roots of $D(Z)$ inside $\partial(R)$ is en by

3) $$\nu(R) = (2\pi i)^{-1} \int_{\partial(R)} \frac{D'(z)}{D(z)} dz := I_1(R) + I_2(R)$$

$$I_j(R) := (2\pi i)^{-1} \int_{\partial^j(R)} \frac{D'(Z)}{D(z)} dz.$$

are interested in $\lim_{R\to+\infty} \nu(R) = \nu(+\infty)$ which is the number of roots $\text{Re } z \geq 0$.

First consider $I_1(R)$ for large R. Now

$$\frac{D'(z)}{D(z)} - \frac{2}{z} = \frac{zk'(z) - 2k(z)}{z(z^2 + k(z))} .$$

Since $tk_j(t) \varepsilon L^1$, both $k(z)$ and $k'(z)$ are bounded for $\text{Re } z \geq 0$: $|k(z)| \leq M$ $|k'(z)| \leq M$, $\text{Re } z \geq 0$ for some real $M > 0$. Thus for $z \varepsilon \partial^1(R)$

$$\left|\frac{D'(z)}{D(z)} - \frac{2}{z}\right| \leq M \frac{R + 2}{R(R^2 - M)}$$

and hence

$$|I_1(R) - 1| = |I_1(R) - (2\pi i)^{-1} \int_{\partial^1(R)} \frac{2}{z} \, dz| \leq M \frac{R + 2}{2(R^2 - M)} \to 0 \quad \text{as} \quad R \to +\infty,$$

or in other words $I_1(+\infty) = 1$.

Next consider $I_2(R)$ for large R. Now

$$(4.4) \qquad I_2(R) = (2\pi i)^{-1} \int_R^{-R} D'(iy)/D(iy) i dy = (2\pi)^{-1}(\arg D(-iR) - \arg D(iR)).$$

Here we use the principal branch $|\arg z| < \pi$ of the logarithm. Since $D(iR) = -R^2 + k(iR)$ it is easily seen that $D(-iR) = \overline{D(iR)}$ and consequently tha $\arg D(-iR) = -\arg D(iR)$. As a result we have

$$I_2(R) = -\arg D(iR)/\pi.$$

Now $D(0) = b_1 b_2 > 0$. Moreover $|k^*(iR)| \leq b_1 b_2$ for $R > 0$ so that it is clear that

$$\text{Re } D(iR) \to -\infty \quad \text{as} \quad R \to +\infty$$

$$|\text{Im } D(iR)| \leq b_1 b_2 \quad \text{for} \quad R \geq 0.$$

These facts imply that $\arg D(iR)$ approaches an odd multiple of π as $R \to +\infty$, $\arg D(+i\infty) = (1 - 2m)\pi$ for some $m = 0, \pm 1, \pm 2, \ldots$. Thus, $I_2(+\infty) = 2m - 1$.

It follows now that $\nu(+\infty) = I_1(+\infty) + I_2(+\infty) = 2m$. Since this is the number roots of $D(z)$ with $\text{Re } z \geq 0$ it must be the case that $m \geq 0$. □

On the basis of Theorem 4.1 we say that the delay predator-prey model (4.2') sually" has an unstable equilibrium. We do this on the grounds that: firstly, only one case (out of infinitely many possible cases) in this theorem is the uilibrium A.S. and secondly, as we shall see below, for linear combinations of e generic kernels we have instability. In addition, we have not been able to nstruct kernels for which case (b) holds.

The following corollary shows that any model (4.2') with even a "slight amount delay" in it has an unstable equilibrium.

COROLLARY 4.2 <u>Suppose that in addition to</u> $k_j(t) \varepsilon L^1_+$, $tk_j(t) \varepsilon L^1$, <u>and</u> $_j|_1 = 1$ <u>the delay kernels in</u> (4.2') <u>satisfy</u> $k''_j(t) \geq 0$, $k'_j(t) \leq 0$ <u>for all</u> > 0, <u>and</u> $k_j(\infty) = k'_j(\infty) = 0$. <u>Then the equilibrium of the delay Volterra model</u> .2') <u>is unstable</u>.

Proof. Several integration by parts show that

$$C_j : = \int_0^\infty k_j(t)\cos Rtdt = -R^{-2} \int_0^\infty k''_j(t)(\cos Rt - 1)dt \geq 0$$

$$S_j : = \int_0^\infty k_j(t)\sin Rtdt = -R^{-1} \int_0^\infty k'_j(t)(1 - \cos Rt)dt \geq 0$$

r all $R > 0$. Since $\text{Im } D(iR) = -b_1b_2(C_1S_2 + C_2S_1)$ it follows that $\text{Im } D(iR) < 0$, > 0. This implies (together with the facts that $|\text{Im } D(iR)|$ is bounded, $D(0) =$ $_2 > 0$ and $\text{Re } D(+i\infty) = -\infty$ as in the proof of Theorem 4.1 above) that gD(+i∞) = -π (i.e., $m = 1$ in Theorem 4.1). □

Any "weak" generic kernels $k_j(t) = T_j^{-1} \exp(-t/T_j)$, $T_j > 0$ would serve to illustrate this corollary.

If on the other hand the delay kernels $k_j(t) = T_j^{-2} t \exp(-t/T_j)$, $T_j > 0$ are used in (4.2') in order to model a situation in which there is a genuine delay in the response of both growth rates, then the equilibrium is still unstable. To see this we note that all of the hypotheses on k_j in Theorem 4.1 hold. Furthermore, a straightforward calculation yields

$$D(iR) = (b_1 b_2 A - R^2(A^2 + B^2) - ib_1 b_2 B)/(A^2 + B^2)$$

where $A = (1 - R^2T_1^2)(1 - R^2T_2^2) - 4R^2T_1T_2$ and $B = 2R(T_1 + T_2)(1 - R^2T_1T_2)$ and consequently

$$\operatorname{Im} D(iR) \begin{cases} > 0, & R^2 > 1/T_1T_2 \\ < 0, & R^2 < 1/T_1T_2. \end{cases}$$

Since it turns out that

$$\operatorname{Re} D(iR) = -\frac{(T_1 + T_2)^2 + b_1 b_2 T_1^2 T_2^2}{T_1 T_2 (T_1 + T_2)^2} < 0$$

at the point where $D(iR)$ crosses the real axis; i.e. when $R^2 = 1/T_1T_2$, it follows that $\arg D(+i\infty) = -\pi$ (see FIGURE 4.1).

For Volterra's original delay model (4.2) the proofs of both Theorem 4.1 and Corollary 4.2 can be repeated as stated with $k_2^*(z)$ set formally equal to one. This we state as

THEOREM 4.3 *When the hypotheses on the delay kernel* $k_2(t)$ *are ignored both Theorem* 4.1 *and Corollary* 4.2 *apply to Volterra's original predator-prey model* (4.2) *with* $D(z)$ *redefined as* $D(z) := z^2 + b_1 b_2 k_1^*(z)$.

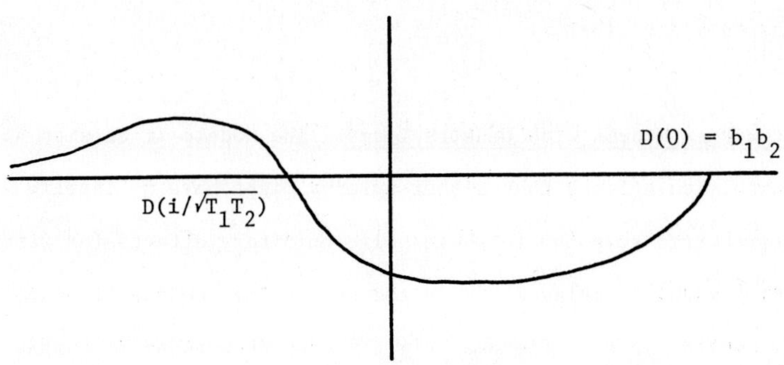

Graph of $D(iR)$, $R > 0$ for the predator-prey model (4.2') with "strong" generic delays in the response to interspecies interactions.

FIGURE 4.1

As an example of the application of this theorem to Volterra's original delay el (4.2) suppose we let the delay in the predator's growth rate response be a ear combination of the two generic delay kernels:

$$k_1(t) = \left(\frac{a}{T} + \frac{bt}{T^2}\right) \exp(-t/T),$$

$$T > 0, \quad a \text{ and } b \geq 0, \quad a + b = 1.$$

s kernel fulfills all the hypotheses in Theorem 4.3 (as stated in Theorem 4.1). this case, $D(z) = z^2 + b_1b_2[a/(zT + 1) + b/(zT + 1)^2]$ and, it turns out,

$$\text{Im } D(iR) = -b_1b_2RT\left(\frac{a}{1 + R^2T^2} + \frac{2b}{(1 + R^2T^2)^2}\right) < 0$$

$R > 0$. As a result $\arg D(+i\infty) = -\pi$ and the equilibrium is unstable. Since s kernel generically covers the case of delayed predator growth rate response ther or not any instantaneous response is present (i.e. whether $a > 0$ or 0) or whether or not the maximum response is genuinely delayed (i.e. whether a or $b \leq a$), we again see that Volterra's delay predator-prey model generically has an unstable equilibrium.

4.2 <u>Predator-Prey Models with Density Terms</u>. The models in Section 4.1 ignore any self-inhibition effects that either species might have on itself. Furthermore, the growth rate response to these self-inhibitory effects (or resource limitation effects) might be delayed, as in the one-species models of Chapter 3, although in this section we will discuss only the case when these responses have no delays, postponing until Section 4.3 the case when such delays are present. Specifically we consider the model

$$(4.5)\qquad \begin{aligned} &\text{(a)} \qquad N_1'/N_1 = b_1(1 - N_1/c) - a_{12}\int_{-\infty}^{t} k_2(t - s)N_2(s)ds \\ &\text{(b)} \qquad N_2'/N_2 = -b_2 + a_{21}\int_{-\infty}^{t} k_1(t - s)N_1(s)ds \end{aligned}$$

under the usual assumptions on the delay kernels $k_j(t)$. This model differs from the Volterra delay model (4.2') by the inclusion of the self-inhibition term $-b_1N_1/c$ which causes N_1 to behave logistically with finite carrying capacity $c > 0$ in the absence of predators. Since such a term generally contributes to th stability of the equilibrium we expect to see a trade-off between its stabilizing influence and the destabilizing influence of the delays in the interaction terms as seen in Section 4.1. We expect then that the equilibrium will be unstable for large c (for a given delay kernel).

Note that the equilibrium of (4.5) is given by

$$(4.6)\qquad e_1 = b_2/a_{21}, \quad e_2 = b_1(c - b_2/a_{21})/ca_{12}.$$

If the carrying capacity $c < b_2/a_{21}$ then $e_2 < 0$ and there is no positive equilibrium. In this case we have the following theorem.

THEOREM 4.4 *Assume that* $k_j(t)\varepsilon L^1_+$, $|k_j|_1 = 1$ *in* (4.5) *and that* $c < b_2/a_{21}$. *If* $N_i(t)$ *is a positive solution of* (4.5) (*and* $N_1(t)$ *has* *nded initial data* $N_1^0(t)$, $t \leq 0$) *then* $N_1(+\infty) = c$ *and* $N_2(+\infty) = 0$.

Thus if the natural carrying capacity c of the prey is too small then the y is incapable of supporting the predators who consequently go to extinction le the prey tend to this carrying capacity. Before proving this theorem we e and prove a lemma.

LEMMA (i) *If* $N(t)$ *is a function defined and bounded for all* t *such that* ∞) *exists and if* $k(t)\varepsilon L^1_+$, $|k|_1 = 1$ *then* $\int_{-\infty}^t k(t-s)N(s)ds \to N(+\infty)$ *as* $+\infty$.

(ii) *If* $0 < c < b_2/a_{21}$ *and* $N_j(t)$ *is a positive solution of* (4.5) *then* *ner* $N_1 \downarrow c$ *as* $t \to +\infty$ *or there exists a first* $t' \geq 0$ *such that* $N_1(t') = c$ *which case* $N_1(t) \leq c$ *for all* $t \geq t'$.

Proof. (i) Given any $\varepsilon > 0$ let $T = T(\varepsilon) > 0$ be so large that for $t \geq T$ $t) - N(+\infty)| \leq \varepsilon$. Then for $t \geq T$

$$|\int_{-\infty}^{t} k(t-s)N(s)ds - N(+\infty)| \leq |\int_{-\infty}^{t} (N(s) - N(+\infty))k(t-s)ds|$$

$$\leq \int_{-\infty}^{T} + \int_{T}^{t} |N(s) - N(+\infty)|k(t-s)ds \leq K\int_{t-T}^{\infty} k(s)ds + \varepsilon \to \varepsilon$$

$t \to +\infty$, which since ε was arbitrary implies (i).

(ii) Suppose first that $N_1(0) > c$. Then for as long as $N_1(t) \geq c$ equation 5a) implies $N_1' \leq 0$ so that either N_1 decreases for all t or there exists a st $t' > 0$ such that $N_1(t') = c$.

Consider the first possibility. It follows that $N_1(+\infty)$ exists. We wish to argue that $N_1(+\infty) = c$. If not, then $N_1'/N_1 < b_1(1 - N_1(+\infty)/c) = \text{constant} < 0$ which implies that $N_1 \to 0$ exponentially, a contradiction. Thus the first possibility is $N_1 \downarrow c$.

Consider now the second possibility. We wish to argue that $N_1(t) \leq c$ for all $t \geq t'$. If this were not the case then there would exist a $t'' > t'$ such that $N_1(t'') > c$, $N_1'(t'') > 0$. But this contradicts (4.5a) at $t = t''$ which implies at $t = t''$ that $0 < N_1'/N_1 \leq b_1(1 - N_1/c) < 0$.

Thus if $N_1(0) > c$ we obtain the two alternatives of the Lemma.

Finally suppose that $N_1(0) \leq c$. Then just as above $N_1(t)$ must be less than or equal to c for all $t \geq 0$ for if not we would obtain a contradiction to (4.5a) at a point $t = t'' > 0$ where $N_1(t'') > c$, $N_1'(t'') > 0$. □

Note that the proof of part (ii) remains applicable to any model in which the interaction terms (which are linear in (4.5)) are nonpositive in the prey equation and nonnegative in the predator equation.

<u>Proof of Theorem</u> 4.4. According to the Lemma, part (ii), above there are two possibilities for N_1.

(i) Suppose $N_1 \downarrow c$ as $t \to +\infty$. Then the right hand side of (4.5b) tends (by the Lemma, part (i)) to $-b_2 + a_{21}c < 0$ so that for large t we have $N_2'/N_2 \leq (-b_2 + a_{21}c)/2 < 0$ which implies the desired result that $N_2(+\infty) = 0$.

(ii) Suppose $N_1 \leq c$ for all $t \geq t' \geq 0$ for some $t' \geq 0$. Then for $t \geq t'$

$$0 \leq \int_{-\infty}^{t'} N_1(s)k_1(t - s)ds \leq K \int_{t-t'}^{\infty} k_1(s)ds \to 0 \quad \text{as} \quad t \to +\infty$$

where K is bound for N_1 for all t. Thus

$$b_2 + a_{21}\int_{-\infty}^{t} N_1(s)k_1(t - s)ds \le -b_2 + a_{21}c + \int_{-\infty}^{t'} N_1(s)k_1(t - s)ds \to -b_2 + a_{21}c < 0$$

$t \to +\infty$ so that for large t, $N_2'/N_2 \le (-b_2 + a_{21}c)/2 < 0$ and again (+∞) = 0.

Finally we must argue in this case that $N_1(+\infty) = c$. Since $0 \le N_1(t) \le c$ for rge t we have $0 \le L = \liminf_{+\infty} N_1 \le S = \limsup_{+\infty} N_1 \le c$. Suppose $L < c$. t $t_n \to +\infty$ be such that $N_1(t_n) \to L$ and $N_1'(t_n) = 0$. Then from (4.5a) we get e contradiction $0 = N_1'(t_n)/N_1(t_n) \le b_1(1 - L/c) < 0$. Thus $L = c$ which implies at $L = S = c = N_1(+\infty)$. □

Next we consider the case when $c > b_2/a_{21}$ and the equilibrium (4.6) is posi- ve. If c is very large so that (4.5) is "close to" the Volterra model (4.2') nsidered in the previous section then we expect the equilibrium to be unstable as s the case for (4.2'). On the other hand if c is close to b_2/a_{21} so that the lf-inhibition is as large as possible (keeping the equilibrium positive of course) might expect its stabilizing influence to stabilize the equilibrium. This is ughly speaking exactly what happens. More specifically we will prove the follow- g theorem for this case when $c > b_2/a_{21}$.

THEOREM 4.5 Assume $k_j(t)\varepsilon L_+^1$ and $tk_j(t)\varepsilon L^1$.

(i) There exists a positive constant $\varepsilon_0 > 0$ such that if the carrying pacity c satisfies $b_2/a_{21} < c < b_2/a_{21} + \varepsilon_0$ then the positive equilibrium .6) of the model (4.5) is (locally) A.S.

(ii) If the hypotheses of Theorem 4.1 hold with $m \neq 0$, then the equilibrium (4.5) is unstable for c sufficiently large.

It is well known that the nondelay version of (4.5) has a globally asymptotic- ly stable equilibrium in the first quadrant (the equilibrium (4.6) if $c > b_2/a_{21}$

and the equilibrium $e_1 = c$, $e_2 = 0$ if $0 < c \leq b_2/a_{21}$). Thus the delay model (4.5) is different from this nondelay version (which is the Lotka-Volterra predator-prey model (4.1) plus a finite carrying capacity for the prey species) only in that case (ii) of Theorem 4.5 arises, that is to say only in that large prey carrying capacities in the presence of delays leads to instabilities.

The three cases described by Theorems 4.4 and 4.5 are graphically illustrated by numerically solved examples in Chapter 6, Section 6.3 below (also see Cushing (1976a)). The transition of stability as c varies through the critical value b_2/a_{21} suggests the possible bifurcation of periodic solutions or limit cycles. The existence of periodic solutions will be discussed in Chapter 6.

<u>Proof of Theorem</u> 4.5. (i) The characteristic equation of the linearized version (at the equilibrium (4.6)) of (4.5) is

$$(4.7) \qquad D(z) := z(z + b_1e_1/c) + e_1e_2a_{12}a_{21}k_1^*(z)k_2^*(z) = 0.$$

Let $\varepsilon = a_{21}/b_2 - 1/c$. Then

$$(4.8) \qquad D(z,\varepsilon) := z(z + b_1) + \varepsilon b_1e_1(-z + b_2k_1^*(z)k_2^*(z)).$$

We will prove part (i) by showing that for $\varepsilon > 0$ small $D(z,\varepsilon)$ has no roots in the right half plane nor on the imaginary axis. Suppose for the purposes of contradiction that this is not true and as a result there exist sequences $\varepsilon_n \to 0$, z_n such that $\text{Re } z_n \geq 0$ and $D(z_n,\varepsilon_n) = 0$. From (4.7) and the fact tha both $k_j^*(z)$ are bounded for $\text{Re } z \geq 0$ follows immediately that the sequence z_n is bounded and consequently we may assume (by extracting a subsequence and relabel ing if necessary) that $z_n \to z_0$ for some z_0, $\text{Re } z_0 \geq 0$. Since $D(z_n,\varepsilon_n) = 0$ w find as $n \to +\infty$ in (4.8) that $z_0(z_0 + b_1) = 0$ and hence $z_0 = 0$. We will estab

h the desired contradiction by showing that the implicit function theorem im-
es that the only zeros of D near 0 for small ε lie in the left half plane.

Clearly $D(0,0) = 0$ and $\partial D(0,0)/\partial z = b_1 > 0$ so that there exists a unique
ution branch of $D(z,\varepsilon) = 0$: $z = z(\varepsilon)$, $z(0) = 0$ for ε small. But an easy
licit differentiation shows that $z'(0) = -e_1 b_2 < 0$ so that $z(\varepsilon)$ has a de-
asing real part near $\varepsilon = 0$. This proves that $\operatorname{Re} z(\varepsilon) < 0$ for $\varepsilon > 0$ small.

(ii) To consider the case when c is large we write

$$D(z) := z^2 + b_1 b_2 k_1^*(z) k_2^*(z) + (1/c)p(z)$$

$$p(z) := b_1 e_1 (z - b_2 k_1^*(z) k_2^*(z)).$$

er the hypotheses on k_1 and k_2 we know that $h(z) := z^2 + b_1 b_2 k_1^*(z) k_2^*(z)$
no roots on the imaginary axis and a finite number of roots (greater than two)
isfying $\operatorname{Re} z > 0$. Let R be so large that all of these roots satisfy
$< R$ so that $m = \min_{\partial(R)} |h(z)| > 0$ where $\partial(R)$ is, as in Section 4.1, the
ndary of the semi-circle $\operatorname{Re} z \geq 0$, $|z| \leq R$. Since $p(z)$ is bounded on $\partial(R)$
may choose c so large that $|(1/c)p(z)| < m$, $z \varepsilon \partial(R)$ and consequently
$z) - h(z)| < |h(z)|$ on $\partial(R)$. Rouche's Theorem then implies that $D(z)$ has
ctly as many roots (counting multiplicities) inside $\partial(R)$ as does $h(z)$ and
ce as many in the right half plane as does $h(z)$. □

4.3 Predator-Prey Models with Response Delays to Resource Limitation. It is
nted out by May (1973) (and literature citing field data is given to support the
ertion) that in at least some predator-prey interactions delays in prey growth
e response to resource limitations are more significant than those present in
responses to interspecies interactions. Thus May (1973, 1974) considers the
el

$$N_1'/N_1 = b_1(1 - \frac{1}{c}\int_{-\infty}^{t} N_1(s)k(t - s)ds) - a_{12}N_2 \tag{4.9}$$
$$N_2'/N_2 = -b_2 + a_{21}N_1.$$

May investigated the linearization of this model for the generic kernel $k(t) = T^{-2}t \exp(-t/T)$. We will discuss his conclusions in Section 4.7. Here we will consider the more general model

$$N_1'/N_1 = b_1(1 - \frac{1}{c}\int_{-\infty}^{t} N_1(s)k_3(t - s)ds) - a_{12}\int_{-\infty}^{t} N_2(s)k_2(t - s)ds \tag{4.10}$$
$$N_2'/N_2 = -b_2 + a_{21}\int_{-\infty}^{t} N_1(s)k_1(t - s)ds$$

which allows for delays in responses to all interactions. Here $k_j \varepsilon L^1_+$, $|k_j|_1 = 1$.

This model has an equilibrium given by (4.6) and hence has a positive equilibrium if and only if $c > b_2/a_{21}$ which we assume is the case. If (4.10) is linearized about this equilibrium one finds that the characteristic equation for the resulting linear system is

$$D(z) := z(z + e_1b_1c^{-1}k_3^*(z)) + e_1e_2a_{12}a_{21}k_1^*(z)k_2^*(z).$$

THEOREM 4.6 *Assume* $k_j \varepsilon L^1_+$, $|k_j|_1 = 1$, $tk_j(t) \varepsilon L^1$.

(i) *If* k_1 *and* k_2 *satisfy the hypotheses of Theorem* 4.1 *with* $m \neq 0$, *then there exists a constant* $c_0 > 0$ *such that* $c > c_0$ *implies that the equilibrium* (4.6) *of* (4.10) *is unstable*.

(ii) *Suppose that* k_3 *is such that* $F(z) := z + b_1k_3^*(z) = 0$ *has no roots satisfying* $\text{Re } z \geq 0$. *Then there exists a constant* $\varepsilon_0 > 0$ *such that* $b_2/a_{21} < c < b_2/a_{21} + \varepsilon_0$ *implies that the equilibrium of* (4.10) *is* (*locally*) *A.S.*

The condition that $F(z)$ have no roots such that $\text{Re}\, z \geq 0$ means that the
ey N_1, which satisfies a delayed logistic in the absence of predators $N_2 \equiv 0$,
an A.S. equilibrium c in the absence of predators.

Theorem 4.6 is in agreement with Rosenzweig (1971) in that it <u>implies that</u> <u>ichment of the prey species</u> (<u>which increases its carrying capacity</u> c) <u>tends</u> <u>destabilize the predator-prey interaction</u>.

Proof. (i) Rewrite $D(z) = p(z) + c^{-1}q(z)$ where $p(z) = z^2 + b_1b_2k(z)$, $) = k_1^*(z)k_2^*(z)$ and $q(z) = b_1b_2a_{21}^{-1}(zk_3^*(z) - b_2k(z))$. According to the assump-ns on k_1 and k_2, $p(z) = 0$ has a finite number of roots in the right half ne $\text{Re}\, z > 0$. Pick $R > 0$ so large that all of these roots satisfy $|z| < R$ that $m = \min_{\partial(R)} |p(z)| > 0$ where $\partial(R)$ is the boundary of the half circle: $z \geq 0$, $|z| \leq R$. Since $q(z)$ is bounded on $\partial(R)$: $|q(z)| \leq M$, we find that

$$|D(z) - p(z)| \leq c^{-1}M < m \leq |p(z)|, \qquad z\varepsilon\partial(R)$$

c large enough. By Rouche's Theorem $D(z)$ has roots inside $\partial(R)$.

(ii) Let $\varepsilon = a_{21}/b_2 - 1/c > 0$ and consider D as a function of ε as l as of z:

$$D(z,\varepsilon) = z^2 + b_1zk_3^*(z) + \varepsilon b_1b_2a_{21}^{-1}(b_2k(z) - zk_3^*(z)).$$

implicit function theorem together with $D(0,0) = 0$, $D_z(0,0) = b_1 > 0$ implies t $D(z,\varepsilon) = 0$ can be solved uniquely for $z = z(\varepsilon)$, $z(0) = 0$ for ε small. implicit differentiation shows that $z'(0) = -b_2^2a_{21}^{-1} < 0$ so that this unique ution branch lies in the left half plane for $\varepsilon > 0$ small. In view of this we whether D can have roots with $\text{Re}\, z \geq 0$ for $\varepsilon > 0$ small.

Suppose no $\varepsilon_0 > 0$ exists as in the statement of the theorem. Then there exist two sequences $\varepsilon_n \to 0$, z_n with $\operatorname{Re} z_n \geq 0$, $D(z_n, \varepsilon_n) = 0$. Since z_n unbounded is incompatible with $D(z_n, \varepsilon_n) = 0$ we may assume without loss in generality that $z_n \to z_0$ where obviously $\operatorname{Re} z_0 \geq 0$. By continuity z_0 satisfies $z_0 F(z_0) = 0$ and hence by assumption it follows that $z_0 = 0$ which leads us to a contradiction to the uniqueness of the solution branch lying in the left half plane found above by the implicit function theorem. □

If no delay is present in the growth rate response of one of the species to interactions with the other (i.e. if either k_1 or k_2 is taken to be δ_0) then it is easy to see that all of the above results and their proofs carry over as given with either k_1^* or k_2^* replaced by 1. If both $k_1 = k_2 = \delta_0$ (as in May's model (4.9)) then $p(z) = z^2 + b_1 b_2$ will have purely imaginary roots and the hypotheses of Theorem 4.1 needed in part (i) of Theorem 4.6 fail to hold. May's model will be briefly discussed in the next Section 4.4.

To illustrate the results in Theorem 4.6 for (4.10) suppose $k_2 = \delta_0$ so that (4.10) reduces to Volterra's delay model (4.2) with an added response delay to resource limitations for the prey species. As the results in Section 4.2 show, for almost any delay kernel k_1 Volterra's model (4.2) is unstable. This fact and Theorem 4.6 (i) imply that for large carrying capacities c the model (4.10) is usually unstable. While on the other hand if the delay represented by k_3 is not "too large" so that, as the results in Chapter 3 show, the delay logistic with this kernel is A.S. then Theorem 4.6 (ii) shows that (4.10) has a (locally) A.S. equilibrium for c near b_2/a_{21}. To be a little more specific in this latter case suppose we let $k_3(t) = T^{-2} t \exp(-t/T)$, $T > 0$. It was shown in Chapter 3, Section 3.1 that the delay logistic with this kernel has a stable equilibrium provided $T < 2/b_1$ (i.e. $F(z)$ in Theorem 4.6 has no roots satisfying $\operatorname{Re} z \geq 0$ provided $T < 2/b_1$) and thus Theorem 4.6 (ii) applies when T satisfies this condition. On the other hand if $T > 2/b_1$ the delay logistic is unstable and we ght expect (4.10) also to be unstable for c near b_2/a_{21}. This is the subject the next theorem.

THEOREM 4.7 Assume $k_j \varepsilon L^1_+$, $|k_j|_1 = 1$, $tk_j(t)\varepsilon L^1$ and that $F(z)$ (as de-ned in Theorem 4.6) has at least one root with $\mathrm{Re}\, z > 0$. Then there exists an > 0 such that $b_2/a_{21} < c < b_2/a_{21} + \varepsilon_0$ implies that (4.10) has an unstable uilibrium.

Proof. $F(z)$ is analytic for $\mathrm{Re}\, z \geq 0$ so that there cannot be an accumula-n point of zeros in the right half plane. Let $\partial(R)$ be the boundary of the gion: $|z| \leq R$, $\mathrm{Re}\, z \geq x_0$ where $R > 0$ and $x_0 \geq 0$ chosen so that the bound- $\partial(R)$ contains no root of $zF(z)$ while at least one root of $F(z)$ lies in-le $\partial(R)$. Let $m = \min_{\partial(R)} |zF(z)|$ so that $m > 0$ and hence for ε small ough

$$|D(z) - zF(z)| = \varepsilon|b_1 b_2 a_{21}(b_2 k(z) - zk_3^*(z))| < m \leq |zF(z)|, \qquad z\varepsilon\partial(R).$$

che's Theorem implies $D(z)$ has a root inside $\partial(R)$. □

4.4 Stability and Vegetation-Herbivore-Carnivore Systems. One interesting uation arises when the delay logistic for the prey species is unstable and hence is the equilibrium for the general delay predator-prey model (4.10) both for ge c and for c near b_2/a_{21} (Theorems 4.6 (i) and 4.7). It is not neces-ily true in this case that (4.10) has an unstable equilibrium for all prey erent carrying capacities c. The possibility that (4.10) is stable for inter-iate values of c raises an interesting point relative to the controversy ling with the question of whether a carnivore is a necessary stabilizing influ-e in a vegetation-herbivore-carnivore community or whether the limited (vegetation) resources available for the prey are the essential stabilizing influence, the carnivore in this case merely being an "undesirable pest" whose presence serves only to decrease the prey's equilibrium state. (Concerning this debate see Hairston et al. (1960), Slobodkin et al. (1967), Murdoch (1966), Ehrlich and Birch (1967), and May (1973).)

We have already seen in the preceding Section 4.3 that it is possible for a stable vegetation-herbivore (prey) system to be destabilized by the introduction of carnivores (predators) (cf. Theorem 4.6 (i)). None of the results of Section 4.3 imply that an unstable vegetation-herbivore system can be stabilized by the introduction of a carnivore, but neither do they preclude this from happening at least for appropriate values of the inherent carrying capacity c. To see that this can in fact occur we will examine a specific model.

Consider (4.10) with $k_3(t) = T_3^{-2} t \exp(-t/T_3)$, $T_3 > 0$, $k_1(t) = T_1^{-1} \exp(-t/T_1)$, $T_1 > 0$ and $k_2(t) = \delta_0(t)$. The resulting model (which is a generalization of May's model as described at the beginning of Section 4.3 in that it allows for delays in the predator growth rate response to predator-prey interactions) is Volterra's original delay predator-prey model (4.2) with an added delay logistic term for the prey. For these chosen delay kernels we know that (4.10) has an unstable equilibrium for c large and for c near (but greater than) b_2/a_{21} provided $T_3 > 2/b_1$ (cf. Section 3.1).

The characteristic function for this example turns out to be $D(z) = N(z)/(zT_3 + 1)^2(zT_1 + 1)$ where

$$N(z) := (T_1T_3^2)z^5 + (2T_1T_3 + T_3^2)z^4 + (T_1 + 2T_3)z^3$$

$$+ (1 + \alpha T_1 + \beta T_3^2)z^2 + (\alpha + 2\beta T_3)z + \beta$$

$$\alpha = e_1b_1/c = b_1b_2/ca_{21}, \quad \beta = e_1e_2a_{21}a_{12} = b_1b_2(1 - b_2/a_{21}c).$$

wish to show that at least for some ranges of parameter values the equation z) = 0, or what amounts to the same thing the equation $N(z) = 0$, has no ots for which $\operatorname{Re} z \geq 0$. Since $N(z)$ is a polynomial the Hurwitz criteria n be applied to this question. If we consider the case when carnivore response lays are smaller than the delays in the vegetation-herbivore system, then T_1 ll be small. Straightforward calculations show that the five Hurwitzian deter- nants of $N(z)$ (all of whose coefficients are obviously positive) are $= 2T_1T_3 + T_3^2 > 0$ and

$$H_2 = 2T_3^3 + O(T_1), \quad H_3 = T_3^3(2 - T_3\alpha) + O(T_1)$$

$$H_4 = \alpha T_3^3(2 - \alpha T_3 - 2\beta T_3^2) + O(T_1), \quad H_5 = \beta H_4.$$

r all of these to be positive for small T_1 we need $2 - T_3\alpha > 0$ and $- \alpha T_3 - 2\beta T_3^2 > 0$ which when related to the original parameters require

(a) $$c > (b_2/a_{21})(b_1T_3/2)$$

.11)

(b) $$2a_{21}(1 - T_3^2b_1b_2)c > b_1b_2T_3(1 - 2T_3b_2).$$

asmuch as we have assumed $T_3 > 2/b_1$ we see that (4.11a) requires, as expected, at c stay away from b_2/a_{21}. Inequality (4.11b) may or may not serve to fur- er constrain c depending of course on the (relative) signs of the parenthetical pressions. (For one possibility, namely $1 - T_3^2b_1b_2 > 0$, (4.11) is satisfied r all large c. This does not contradict Theorem 4.6 (i) since we also demand at T_1 be small...but how small depends on c as the expression for H_4 ows.)

We conclude then that <u>under certain circumstances an unstable vegetation-</u>

herbivore (prey) system is stabilized by the introduction of a carnivore (predator) even when the herbivore-carnivore (prey-predator) system with unlimited resources (vegetation) is unstable.

These conclusions were put forth by May (1973, 1974) for the simpler model (4.9). We have found it convenient to study this question using c as the crucial parameter, although it is clear that we could instead have used T_3 (and/or T_1) as does May. Even though the discussion centered on the simplistic model (4.10) we again point out that the linearization procedure used to make these conclusions is valid for any model which has the same characteristic equation (i.e. the same linearization).

4.5 Some Other Delay Predator-Prey Models. Although, as we have repeatedly stated, the results of the preceding sections are given for certain classical predator-prey models with quadratic interactions only (i.e. for which the response functions f_i in the general model (1.1) are linear), the linearization method, being local in nature, of course applies to any nonlinear model and the results above (except Theorem 4.4) apply per se to any model with the same linearizations (with of course possibly different interpretations of the parameters). A great many predator-prey models have been proposed and studied which, in one way or another, improve upon or at least differ from these basic quadratic models. (For example six sample models are discussed in Rosensweig (1971). Also see May (1974a, p. 79).) Any of these could be modified to include delayed responses by the inclusion of a Volterra integral of the type we have been considering. In this section we will briefly discuss a few details of several delay predator-prey models which are not of the form of those discussed in the previous sections.

(1) In a long paper Holling (1965) forcefully argues that the response of a predator should not be a linear function of prey density (as it is in the Lotka-Volterra model (4.1)) but should be an S-shaped curve representing a monotonic

crease to a finite saturation level. One possible modification in the Lotka-lterra model (4.1) suggested by Holling would be to replace the term $a_{21}N_1$ in e predator equation by a constant multiple of $d_1N_1(t)/(1 + d_1N_1(t))$. In lling's derivation this latter expression represents the number of "attacks" by edators on prey at time t. If one follows Volterra's derivation of his delay del (4.2) with this expression for the number of attacks in place of simply a ltiple of $N_1(s)$, then one obtains $a_{21}\int_{-\infty}^{t} k(t - s)d_1N_1(s)/(1 + d_1N_1(s))ds$ in ace of the integral in the second equation of (4.2). Thus the per unit growth te of predators would be a function of attacks at all earlier times as weighted the delay kernel.

If we also assume that the per unit growth rate of prey is a function of all st attacks by predators and that this response is S-shaped we obtain the following model

$$
\begin{aligned}
N_1'/N_1 &= b_1(1 - c^{-1}\int_{-\infty}^{t} N_1(s)k_3(t - s)ds) - a_{12}\int_{-\infty}^{t} \frac{N_2(s)}{1 + N_2(s)} k_2(t - s)ds \\
N_2'/N_2 &= -b_2 + a_{21}\int_{-\infty}^{t} \frac{N_1(s)}{1 + N_1(s)} k_1(t - s)ds.
\end{aligned} \tag{.12}
$$

re we have assumed that in the absence of predator the prey population is govned by a delay logistic. We have also assumed that N_i is measured in units ich made $d_i = 1$.

Model (4.12) has an equilibrium in the right half plane, provided $a_{21} > b_2$:

$$
e_1 = \frac{b_2}{a_{21} - b_2}, \qquad e_2 = \frac{b_1(c - e_1)}{a_{12}c - b_1(c - e_1)}. \tag{.13}
$$

ere are several cases to be considered depending on relative values of certain rameters. We will very briefly consider each in turn.

(a) Assume $a_{21} > b_2$ so that $e_1 > 0$. If $c < e_1$ then $e_2 < 0$ and the arguments proving Theorem 4.4 carry over to (4.12) (almost verbatim) at least when $k_3 = \delta_0$. Thus <u>if</u> $a_{21} > b_2$ <u>all positive solutions (with bounded initial data) satisfy</u> $N_1(+\infty) = c$, $N_2(+\infty) = 0$ <u>when</u> $k_3 = \delta_0$ (i.e., <u>when prey growth rate response to resource limitation is instantaneous</u>) <u>provided the inherent prey carrying capacity is small, namely provided</u> $c < e_1$.

Next consider the case when (4.13) lies in the first quadrant, i.e. the case

$$(4.14) \qquad c > e_1 \quad \text{and} \quad c(a_{12} - b_1) + b_1 e_1 > 0.$$

The characteristic function of the linearized model at the equilibrium (4.13) under these conditions is

$$D(z) := z(z + \alpha k_3^*(z)) + \beta k(z), \qquad k(z) := k_1^*(z) k_2^*(z)$$

$$\alpha = e_1 b_1 / c, \qquad \beta = e_1 e_2 a_{12} a_{21} (1 + e_1)^{-2} (1 + e_2)^{-2}.$$

(i) First of all, as $c \downarrow e_1$ we see that $D(z)$ is close to $z(z + b_1 k_3^*(z))$ and just as in the proofs of Theorems 4.6 (ii) and 4.7 we can argue that <u>if</u> $a_{21} > b_2$ <u>and the inherent prey carrying capacity satisfies</u> $c > e_1$, <u>yet is close to</u> e_1 <u>and such is that</u> (4.14) <u>holds, then the equilibrium</u> (4.13) <u>of the model</u> (4.12) <u>is (locally) A.S. or is unstable according to whether the delay logistic for the prey is (locally) A.S. or unstable respectively</u>. Here of course the usual assumptions concerning the delay kernels are made: $k_j \varepsilon L_+^1$, $|k_j|_1 = 1$ and $t k_j(t) \varepsilon L^1$. Thus, for c near the critical value e_1 the model (4.12) behaves for c near b_2/a_{21} as does the model (4.10) with linear response functions.

Does (4.12) still behave like (4.10) for large c, i.e. is (4.12) unstable for large c? The answer is "yes" provided (4.14) allows c to be large. Thus we distinguish two cases: $a_{12} > b_1$ and $a_{12} < b_1$.

(ii) Suppose $a_{12} > b_1$. Then (4.14) holds (and $e_1 > 0$) for all large

In this case $D(z)$ becomes close to $z^2 + \beta k(z)$,

$= b_1 b_2 (a_{12} - b_1)(a_{21} - b_2) a_{12}^{-1} a_{21}^{-1}$ and hence (with β replacing $b_1 b_2$) the proof

Theorem 4.6 applies: *if* $a_{21} > b_2$, $a_{12} > b_1$ *and if* k_1, k_2 *satisfy the potheses of Theorem* 4.1 *with* $m \neq 0$ *then* (4.12) *has an unstable positive equilib-um* (4.13) *for large* c.

(iii) Suppose $0 < a_{12} < b_1$ (but still $a_{21} > b_2$). Then (4.14) only holds nd hence $e_1 > 0$ only for) $c \varepsilon (e_1, b_1 e_1/(b_1 - a_{12}))$. Since we know what happens $c \downarrow e_1$ (see (i)) we consider the case when $c \uparrow b_1 e_1/(b_1 - a_{12})$, which im-ies $e_2 \to +\infty$ and $D(z)$ gets close to $z(z + (b_1 - a_{12})k^*(z))$. Thus from only ight modifications in the proofs of Theorems 4.6 (ii) and 4.7 we conclude that *if* $_1 > b_2$, $0 < a_{12} < b_1$ *and* c *is less than, but close to* $c_0 := b_1 e_1/(b_1 - a_{12})$ *en* (4.12) *has a* (*locally*) *A.S. or an unstable equilibrium* (4.13) *if* $z) := z + (b_1 - a_{12})k^*(z)$ *has no roots* $\text{Re } z \geq 0$ *or at least one root* $\text{Re } z > 0$ *spectively*.

We see that it is essentially only in the possibility of case (iii) that the edator-prey model (4.12) with Holling response functions differs from the quad-tic delay model (4.10) with linear response functions. In this case (iii), when e inherent predator death rate b_2 is small enough and the inherent prey birth te b_1 is large enough, in order to have a positive equilibrium at all it is cessary that the prey's inherent carrying capacity c be restricted to a finite terval (e_1, c_0); moreover, as c approaches its upper bound c_0 the predator uilibrium e_2 grows without bound and the stability or instability of the sys-m's equilibrium depends on the prey's delayed response to its own resource limi-tion (i.e., on k_3). As an example, if no logistic delay is present, $k_3 = \delta_0$, en $F(z) = z + b_1 - a_{12}$, $b_1 > a_{12}$ has one negative real root so that the uilibrium is (locally) A.S. for c both near e_1 and near c_0. Here we have a tuation where unlike the previous cases enrichment of the prey does not lead to an unstable equilibrium (at least if c is not increased beyond c_0). On the other hand if $k_3(t) = T^{-2}t \exp(-t/T)$ with $T > 2/(b_1 - a_{12})$ then the equilibrium is stable for c near e_1 and unstable for c near c_0.

(b) If $0 < a_{21} < b_2$ so that $e_1 < 0$ then from (4.12) we have that $N_2'/N_2 \leq -b_2 + a_{21} < 0$ and hence $N_2 \downarrow 0$ exponentially as $t \to +\infty$. Thus it shoul follow from (4.12) that N_1 behaves according to the delayed logistic. For example if $k_3 = \delta_0$ then as in Section 4.2 we can prove that $N_1(\infty) = c$.

(2) The Leslie predator-prey model (see J. M. Smith (1974) for a discussion of this model) differs from all of the predator-prey models studied above in that the predator's response function is assumed to be basically logistic except that its carrying capacity is a function of the prey density (in fact it is assumed to be proportional to N_1). This model

$$N_1'/N_1 = b_1(1 - c^{-1}N_1) - a_{12}N_2$$

$$N_2'/N_2 = b_2(1 - N_2/a_{21}N_1)$$

is easily shown to have a unique, positive A.S. equilibrium

$$e_1 = b_1c/(b_1 + ca_{12}a_{21}), \quad e_2 = a_{21}b_1c/(b_1 + ca_{12}a_{21})$$

for all values of the (positive) parameters in the model. If in addition to delays in the prey's response to its resource limitation and to increased predator density as assumed in earlier models, we also assume that the predator's reaction to its resource limitation (prey) is delayed and that its "carrying capacity" depends in a delayed manner on prey density, we obtain the following delay version of the above Leslie model

$$N_1'/N_1 = b_1(1 - c^{-1}\int_0^\infty N_1(t - s)k_{11}(s)ds) - a_{12}\int_0^\infty N_2(t - s)k_{12}(s)ds$$

$$N_2'/N_2 = b_2(1 - (\int_0^\infty N_2(t - s)k_{22}(s)ds)/(a_{21}\int_0^\infty N_1(t - s)k_{21}(s)ds))$$

$$k_{ij}\varepsilon L^1_+, \quad |k_{ij}|_1 = 1$$

ich has the same positive equilibrium as the above nondelay model. (Similar lay expressions were used and more elaborately derived in a two predator, one ey model by Caswell (1972).) The characteristic equation of the linearized rsion of this model is

$$D(z): = (z + e_1b_1c^{-1}k^*_{11}(z))(z + b_2k^*_{22}(z)) + e_2b_2a_{12}k^*_{12}(z)k^*_{21}(z) = 0.$$

Rather than study this equation in a detailed manner such as in Sections - 4.3 above we confine ourselves to a few simple, rough observations. If > 0 is small then $D(z)$ is nearly $(z + b_1k^*_{11})(z + b_2k^*_{22})$ whose roots are of urse those of each factor. Each factor is the characteristic function of a deyed logistic. Thus if these delay kernels are such that these logistic models e A.S. then we expect the delay Leslie model to be A.S. for small prey carrying pacity c. If, on the other hand, at least one of these delayed logistics is stable then we expect instability in the Leslie model for small c.

For large c the characteristic equation is nearly

z): $= z(z + b_2k^*_{22}) + b_1b_2k^*_{12}k^*_{21} = 0$, an equation of the same form as that of e model studied in Section 4.3. For example, if b_2 and k_{22} are such that z): $= z + b_2k^*_{22}(z)$ has no roots with $\operatorname{Re} z \geq 0$, then $p(z)$ has no roots with $z \geq 0$ for b_1 small (cf. the proof of Theorem 4.6 (ii)). Thus, under these nditions $D(z)$ has no roots with $\operatorname{Re} z \geq 0$ for large c. As a result the lay Leslie model does not necessarily predict an unstable equilibrium due to prey enrichment (increased c).

For some analysis of a delay model similar to the above delay Leslie model, but with a Holling prey response function see Mac Donald (1976).

(3) In his experimental studies of the organism Isochrysis galbana in a nitrate limited chemostat Caperon (1968) derives and utilizes a delay model of integrodifferential equations which govern the growth rates of the nutriant concentration N_1 and the population concentration N_2 in the growth chamber. His model has the form

$$N_1' = -a(c + N_2)f(N_1) + 1/N_1$$

$$N_2'/N_2 = -b + af(N_1) \tag{4.15}$$

$$f(N_1) := \frac{\int_{-\infty}^{t} N_1(s)k(t - s)ds}{1 + \int_{-\infty}^{t} N_1(s)k(t - s)ds}, \quad k \varepsilon L_+^1, \quad |k|_1 = 1$$

where $a,b,c > 0$ are certain positive, physical constants which we will not define here. (We have scaled N_i in order to eliminate two parameters from the model as given in Caperon (1968).)

This model has equilibrium

$$e_1 = b(a - b)^{-1}, \quad e_2 = (a - b - b^2c)b^{-2}$$

which is positive if and only if $a - b > b^2c$, which we assume holds.

The characteristic function of the linearized system is

$$D(z) := z(z + \alpha + \delta k^*(z)) + \gamma k^*(z)$$

$$\delta = a(c + e_2)(1 + e_1)^{-2} > 0, \qquad \alpha = e_1^{-2} > 0, \qquad \gamma = a^2 e_2 (1 + e_1)^{-3} > 0.$$

Caperon used a "block" delay kernel which we might smooth out and approximate ith our standard, generic delay kernel $k(t) = T^{-2} t \exp(-t/T)$, $T > 0$. For this ernel $D(z) = N(z)/(zT + 1)^2$ where

$$N(z) := T^2 z^4 + T(2 + \alpha T) z^3 + (1 + 2\alpha T) z^2 + (\alpha + \delta) z + \gamma.$$

l of the coefficients of $N(z)$ are positive and the Hurwitzian determinants rn out to be $H_1 = T(2 + \alpha T) > 0$, $H_4 = \gamma H_3$ and

$$H_2 = T[2\alpha^2 T^2 + (4\alpha - \delta)T + 2]$$

$$H_3 = T[-\gamma\alpha^2 b^2 T^3 + 2\alpha(\alpha^2 + \alpha\delta - 2\gamma\alpha)T^2 + ((\alpha + \delta)(4\alpha - \delta) - 4\gamma)T + 2(\alpha + \delta)].$$

ince $H_2 > 0$, $H_3 > 0$ for T small and $H_3 < 0$ for T large we conclude that aperon's model (4.15) has an A.S. equilibrium for small delays T and an unstable quilibrium for large delays T.

(4) In a frequently referenced paper Wangersky and Cunningham (1957) introuced a delay predator-prey model derived from the classical Lotka-Volterra model 4.1) by replacing the predator response term $a_{21}N_1N_2$ by the lagged terms $_{21}N_1(t - T)N_2(t - T)$, $T > 0$. A more realistic delay version of their model would hen be

$$N_1'/N_1 = b_1(1 - c^{-1}N_1) - a_{12}N_2$$

4.16)

$$N_2' = -b_2 N_2 + a_{21} \int_{-\infty}^{t} N_2(s) N_1(s) k(t - s)\, ds$$

where (as in Knolle (1976)) we have included a finite carrying capacity c for the prey in the absence of predators.

This model has the same equilibrium (4.6) as that of the similar model (4.5). This equilibrium is positive if $c > b_2/a_{21}$ and the characteristic function of the linearization is

$$D(z) := (z + \alpha)(z + b_2 - b_2 k^*(z)) + \beta k^*(z)$$

$$\alpha = b_1 b_2/ca_{21} > 0, \qquad \beta = b_1 b_2(1 - b_2/ca_{21}) > 0.$$

Rather than study this characteristic function in any general setting for arbitrary delay kernels, let us take $k(t) = T^{-2} t \exp(-t/T)$, $T > 0$, the generic delay kernel with delay T. In this case (4.16) is a "smoothed out" version of the original model of Cunningham and Wangersky. We find then that $D(z) = N(z)/(Tz + 1)^2$ where

$$N(z) := T^2 z^4 + T(2 + T(\alpha + b_2))z^3 + (1 + 2(\alpha + b_2)T + \alpha b_2 T^2)z^2 + \alpha(1 + 2Tb_2)z + \beta.$$

It turns out that the first two H_1 and H_2 of the four Hurwitzian determinants of $N(z)$ are positive for all values of the parameters. Since $H_4 = \beta H_3$ the stability of the equilibrium reduces to the sign of

$$H_3 = \alpha(1 + 2Tb_2)H_2 - \beta T^2(2 + T(\alpha + b_2))^2.$$

We distinguish two cases: large inherent prey carrying capacity c and c close to the critical value b_2/a_{21}. For c large, $H_3 \sim -b_1 b_2 T^2(2 + Tb_2)^2 < 0$ and we have instability. For $c \sim b_2/a_{21}$ we find that $H_3 \sim b_1(1 + 2Tb_2)H_2 > 0$ and we have A.S.

Thus, the generalized model (4.16) of Cunningham and Wangersky has, for any "strong" generic delay kernel $k(t) = T^{-2}t \exp(-t/T)$, $T > 0$, an unstable equilibrium for large inherent prey carrying capacity c and an A.S. equilibrium if $c \sim b_2/a_{21}$, $c > b_2/a_{21}$.

4.6 The Stabilization of Predator-Prey Interactions. It has often been pointed out that it is difficult to obtain a sustained predator-prey interaction in a laboratory experiment (see e.g. Gause (1934), Huffaker (1958)). It seems that it is necessary to control externally some parameter in a repeated or continuous manner to obtain the coexistence of predator and prey. It is also frequently pointed out that time delays must be taken into account in order to explain the oscillations (unstable or stable) which are observed in these experiments (J. M. Smith (1974, p. 33), Caswell (1972), F. Smith (1963)). This is of course consistent with the general trend of results in Sections 4.1 - 4.5 in that time delays tend to destabilize predator-prey interactions, especially for large inherent prey carrying capacities.

Many of our results above are also consistent with some of the procedures used by experimenters which stabilize or at least tend to stabilize the interaction. For example, Luckinbill (1973) obtained the coexistence of *Paramecium aurelia* (prey) and *Didinium nasutum* (predator) by in effect increasing the predator's inherent death rate b_2 and/or decreasing the food available for prey and hence decreasing c (see J. M. Smith (1974, p. 33) for a discussion of this experiment and its relationship to Volterra models). As seen in the above results in Sections 4.2 - 4.3 both of these changes (which tend to make c closer to b_2/a_{21}) tend to stabilize the model (if the delay is unchanged). The delay in this experiment was caused by the delay in division of *Didinium* after capture of prey.

Another important aspect of predator-prey interactions which has been found to contribute to their stability is the possibility that at least some of the prey can find refuges from predators (J. M. Smith (1974), Caswell (1972), Gause (1934), Huffaker (1958)). Suppose that a certain number of prey $g(N_1)(t) \geq 0$ can find cover where they are inaccessible to predators. Thus at any time only $N_1 - g(N_1)$ of the prey are available for contact with predators. If the mixed quadratic terms in the model (4.6)(after multiplication by N_i) are interpreted as describing the responses of growth rates due to inter-species contacts then this model becomes

$$N_1' = b_1(1 - N_1/c)N_1 - a_{12}(N_1 - g(N_1))N_2$$

$$N_2' = -b_2N_2 + a_{21}N_2 \int_{-\infty}^{t} (N_1(s) - g(N_1)(s))k_1(t - s)ds.$$

Here we have assumed for simplicity that the only delays occur in predator responses to prey densities. We will only consider two cases (following J. M. Smith (1974, p. 25)): either $g(N_1) \equiv g_0$, a constant, or $g(N_1) \equiv g_0N_1$ for a constant $0 < g_0 < 1$.

If $g(N_1) \equiv g_0N_1$ (that is, if there is always a fixed fraction g_0 of the prey population under cover) then we find that the above model reduces to (4.6) with a_{12}, a_{21} replaced by $a_{12}(1 - g_0)$, $a_{21}(1 - g_0)$. Thus, since we have in effect decreased a_{21} by introducing refuges in this manner, we see that (all other parameters held fixed) stability has been enhanced. This is because if c is greatly larger than b_2/a_{21} (causing instability according to Section 4.2) then decreasing a_{21} causes b_2/a_{21} to increase towards c thereby promoting stability (again see Section 4.2). In this way providing cover for prey enhances the stability of the model.

Suppose that $g(N_1) \equiv g_0 > 0$, a constant. This is the same as assuming that there is a finite amount of cover for prey which is always utilized by the prey.

other interpretation is that there is a threshold level below which the predator nores (or cannot find) prey (Caswell (1972)). In this case the model becomes

$$N_1' = b_1(1 - N_1/c)N_1 - a_{12}(N_1 - g_0)N_2$$

$$N_2' = -(b_2 + a_{21}g_0)N_2 + a_{21}N_2 \int_{-\infty}^{t} N_1(s)k_1(t - s)ds.$$

us we have in effect increased the predators inherent death rate from b_2 to $+ a_{21}g_0$ as well as introduced a new linear term $a_{12}g_0N_2$ into the prey equa- on. For simplicity we assume that time t is scaled so that $b_2 = 1$ and that e units for N_1 are such that $a_{21} = 1$. Then the resulting model has equilib- um

$$e_1 = 1 + g_0, \qquad e_2 = b_1(1 + g_0)(c - 1 - g_0)/a_{12}c$$

ich is positive provided $c > 1 + g_0 = e_1$, which we assume holds. A lineariza- on about this equilibrium yields

$$x_1' = b_1(g_0^2 - cg_0 - 1)c^{-1}x_1 - a_{12}x_2$$

$$x_2' = e_2 \int_0^t x_1(s)k_1(t - s)ds.$$

: is not difficult to show algebraically that the coefficient of x_1 in the rst equation is negative since $c > 1 + g_0$. As a result the analysis of the aracteristic equation follows exactly that of the model in Section 4.2. This ields the result that: $c > 1 + g_0$ but close to $1 + g_0$ implies a (locally) S. equilibrium while large c implies an unstable equilibrium.

We see then that this model again predicts the stabilizing influence of refuges for prey. For, all other parameters (including c) fixed, an increase in g_0 makes c closer to $1 + g_0$.

4.7 <u>A General Predator-Prey Model</u>. Consider any predator-prey model

$$N_1'/N_1 = b_1 f_1(N_1, N_2), \quad N_2'/N_2 = -b_2 f_2(N_1) \tag{4.17}$$

where f_i is some functional of its arguments for which $f_1(e_1, e_2) = 0$, $f_2(e_1) = 0$ for some constants $e_i > 0$. Here $b_i > 0$ are constants; b_i is the inherent exponential birth (death) rate of prey (predators) in the absence of all constraints. Suppose that in some reasonable and meaningful manner a measure $T_1 \geq 0$ of the delay present in the first (prey) equation is determined. We also assume that some measure $T_2 \geq 0$ is determined for the delay in the second (predator) equation and finally that the T_i are averaged or otherwise combined to yield a measure $T > 0$ of the delay present in the system (4.17). If we make T the unit of time by letting $t^* = t/T$ in (4.17) then this system reduces to one of the same form with b_i replaced by the dimensionless parameters Tb_i:

$$N_1'/N_1 = b_1 T f_1(N_1, N_2), \quad N_2'/N_2 = -b_2 T f_2(N_1) \tag{4.18}$$

where for simplicity we will relabel t^* as t. Finally we must make some assumptions on f_i, at least near the equilibrium $N_i = e_i$, so that (4.18) reflects the predator-prey nature of the interactions, say: $\partial f_1/\partial N_1 \leq 0$ (resource limitation for prey), $\partial f_1/\partial N_2 < 0$ (predator increase causes a decrease in prey growth rate) and $\partial f_2/\partial N_1 < 0$ (prey increase causes an increase in predator growth rate) with suitable definitions of these partial derivatives (say, Fréchet derivatives). More precisely what we assume is that (4.18) is linearizable at $N_i = e_i$ to a system of the form

$$x_1' = -\beta_1(c_{11}\int_0^t x_1(s)k_{11}(t - s)ds + c_{12}\int_0^t x_2(s)k_{12}(t - s)ds)$$

$$x_2' = \beta_2 c_{21}\int_0^t x_1(s)k_{21}(t - s)ds, \qquad \beta_i = b_i T$$

or kernels $k_{ij}\varepsilon L_+^1$, $tk_{ij}\varepsilon L^1$, $|k_{ij}|_1 = 1$ and constants $c_{ij} \geq 0$. This system has characteristic equation

$$D(z): = z(z + \beta_1 c_{11}k_{11}^*(z)) + \beta_1\beta_2 c_{12}c_{21}k_{12}^*(z)k_{21}^*(z) = 0.$$

Suppose β_1 is chosen so small: $0 < \beta_1 < \beta_1^0$ that $z + \beta_1 c_{11}k_{11}^*(z) \neq 0$ or $\mathrm{Re}\, z \geq 0$ (cf. Section 3.4). Arguing essentially as in the proof of Theorem .6 (ii) we can prove that if β_2 is small: $0 < \beta_2 < \beta_2^0$ then $D(z)$ has no oots with $\mathrm{Re}\, z \geq 0$.

Thus, for the general model described above the equilibrium (which is not ecessarily unique) is (locally) A.S. for $b_i T$ sufficiently small (Cushing 1977)).

On the other hand, if $z + \beta_1 c_{11}k_{11}^*(z) = 0$ has at least one root with e $z > 0$ then for $b_2 T$ sufficiently small the equilibrium $N_i \equiv e_i$ of (4.17) is nstable. This can be proved essentially as in the proof of Theorem 4.7.

The following is a generalization of Theorem 4.1 which can be used to inves- igate cases not covered by the preceding statements.

THEOREM 4.8 *Suppose* $k_{ij}\varepsilon L_+^1$, $|k_{ij}|_1 = 1$ *and* $tk_{ij}(t)\varepsilon L^1$. *Assume*

(i) $D(iR) \neq 0$ *for all* $R > 0$

(ii) $c_{11}\frac{d}{dz}k_{11}^*\Big|_{z = R\exp(i\theta)} \to 0$ *uniformly for* $-\pi/2 \leq \theta \leq \pi/2$ *as* $R \to +\infty$.

hen $\arg D(+i\infty) = (1 - 2m)\pi$ *for some* $m = 0, 1, 2, \ldots$ *and the equilibrium of he general predator-prey model* (4.18) *as described above is A.S. if* $m = 0$ *and*

<u>unstable</u> <u>if</u> $m \geq 1$.

Proof. The only modification of the proof of Theorem 4.1 that is needed in order to prove this theorem is that D'/D needs to be compared to $h(z) := 2/z + \beta_1 c_{11} k^{*\prime}_{11}(z)/z$, instead of to $2/z$, for $z \varepsilon \partial^1(R)$ where $k^{*\prime}(z) = (d/dz)(k^*)$. With only a few added details it is easily shown that $|I_1(R) - (2\pi i)^{-1} \int_{\partial^1(R)} h(z)dz| \to 0$ as $R \to +\infty$. But using (ii) one can show that $(2\pi i)^{-1} \int_{\partial^1(R)} h(z)dz \to 1$ as $R \to +\infty$. (More specifically (ii) implies

$$|(2\pi i)^{-1} \int_{\partial^1(R)} c_{11}(k^{*\prime}_{11}(z)/z)dz| \leq (2\pi)^{-1} c_{11} \int_{-\pi/2}^{\pi/2} |k^{*\prime}_{11}(Re^{i\theta})| d\theta \to 0 \quad \text{as} \quad R \to +\infty.$$

$\square$

Examples of kernels which satisfy (ii) are $T^{-2} t \exp(-t/T)$ or $T^{-1} \exp(-t/T)$, $T > 0$ (or any linear combination of these two).

As an application of the use of Theorem 4.8 consider May's model (4.9) with $c > b_2/a_{21}$ (so that the model has a positive equilibrium (4.6)). In this model the only delay in the system is that of the prey's growth rate response to its own resource limitation. The characteristic function is ($k = k_{11}$, $c_{11} = e_1/c > 0$, $c_{12} = e_1 a_{12}/b_1$, $c_{21} = e_2 a_{21}/b_2$)

$$D(z) = z(z + \beta_1 c_{11} k^*(z)) + \beta_1 \beta_2 c_{12} c_{21}.$$

Then

$$\mathrm{Re}\, D(iR) = -R^2 + R\beta_1 c_{11} S(R) + \beta_1 \beta_2 c_{12} c_{21}$$

$$\mathrm{Im}\, D(iR) = R\beta_1 c_{11} C(R)$$

where $k^*(iR) = C(R) - iS(R)$. For the generic delay kernels (with unit delay) we ve

$$C(R) = \begin{cases} (1 + R^2)^{-1} & \text{for} \quad k(t) = \exp(-t) \\ (1 - R^2)(1 + R^2)^{-2} & \text{for} \quad k(t) = t \exp(-t). \end{cases}$$

Thus, for the "weak" generic kernel $k(t) = \exp(-t)$ *May's model (4.9) has A.S. equilibrium for all values of the parameters* $(c > b_2/a_{21})$. This is cause $\text{Im}\, D(iR) > 0$ for all $R > 0$ which implies $\arg D(+i\infty) = \pi$ or $m = 0$ Theorem 4.8. (This argument actually applies to any decreasing, convex kernel in Corollaries 3.5 and 4.2.)

On the other hand, suppose $k(t)$ is the "strong" generic kernel t) $= t \exp(-t)$. Then there are two cases illustrated by FIGURE 4.2 in which e two possible graphs of $D(iR)$ are drawn for $R \geq 0$. They are distinguished means of the sign of $D(i)$, the case $D(i) > 0$ yielding instability and the se $D(i) < 0$ yielding A.S. Now $S(R) = 2R(1 + R^2)^{-2}$ and

$$D(i) = \text{Re}\, D(i) = \beta_1\beta_2 c_{12}c_{21} + \beta_1 c_{11}/2 - 1.$$

$$= (b_1 b_2 c_{12} c_{21})T^2 + \frac{1}{2}(b_1 c_{11})T - 1.$$

us, May's model (4.9) has an A.S. or unstable equilibrium for the "strong" neric delay kernel provided $T < T_0$ *or* $T > T_0$ *respectively where* $T_0 > 0$ *is e unique positive root of the quadratic* $D(i)$.

4.8 *Competition and Mutualism.* Consider the system

$$\text{.19)} \qquad N_i'/N_i = b_i(1 - c_{ij}N_j), \quad i \neq j, \quad 1 \leq i,j \leq 2, \quad c_{ij} > 0.$$

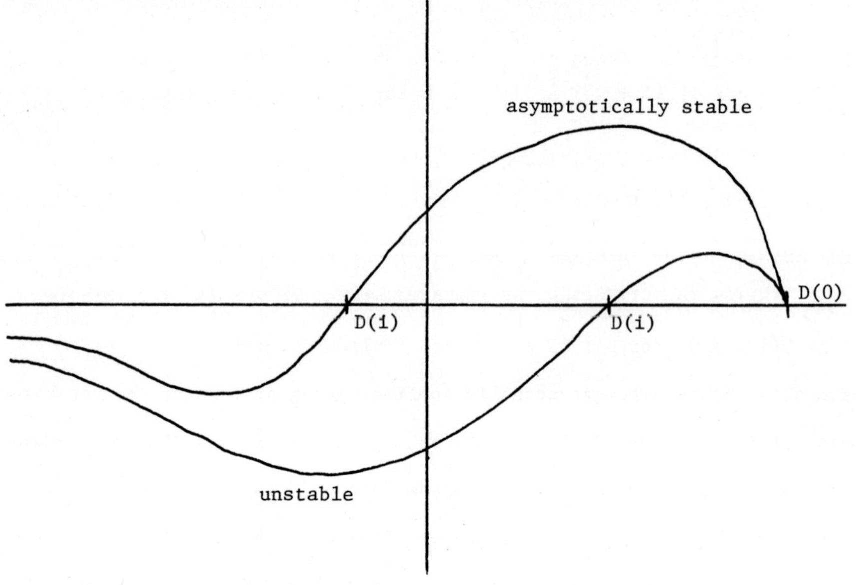

Graph of $D(iR)$, $R > 0$ for May's model (4.9) with a "strong" generic delay kernel.

FIGURE 4.2

Here we assume that b_1, b_2 are nonzero constants of the same sign $b_1 b_2 > 0$ but not necessarily positive. If $b_i > 0$ then (4.19) is perhaps the simplest of models for the interaction of two species who are competing for a common resource: each hampers the others growth rate while each grows exponentially in the absence of the other. If on the other hand $b_i < 0$ then (4.12) would represent two species in a mutualistic interaction both of whom die exponentially in the absence of the other and both of whom aid the others growth rate. It is easy to see (e.g. by investigating the phase plane direction field) that the positive equilibrium

$$e_1 = 1/c_{21}, \quad e_2 = 1/c_{12} \tag{4.20}$$

(4.19) is unstable. We wish here to see what effect the inclusion of time lays in the interaction terms in (4.19) has and specifically to see whether or t such delays could conceivably result in a stable equilibrium. The answer, as will turn out, is essentially that they cannot.

We consider a general model

.21) $$N_i'/N_i = b_i f_i(N_j), \qquad i \neq j, \quad 1 \leq i,j \leq 2, \quad b_1 b_2 > 0$$

ere f_i is a functional satisfying the following conditions: $f_i(e_j) = 0$ for me constants $e_j > 0$, $\partial f_i(e_j)/\partial N_j < 0$ and f_i is sufficiently smooth near $\equiv e_j$ so that (4.21) has a linearization

.22) $$x_i' = -b_i a_i \int_0^t x_j(s) k_i(t - s)ds, \quad i \neq j, \quad 1 \leq i,j \leq 2, \quad a_i > 0$$

th $k_i \varepsilon L_+^1$, $|k_i|_1 = 1$. An example of such a system is the following delay veron of (4.19)

.23) $$N_i'/N_i = b_i(1 - c_{ij} \int_{-\infty}^t N_j(s) k_i(t - s)ds), \quad c_{ij} > 0, \quad b_1 b_2 > 0.$$

The linear system (4.22) has characteristic equation

.24) $$D(z): = z^2 - b_1 b_2 k_1^*(z) k_2^*(z)$$

ich, except for the sign change, is the same as that for the Lotka-Volterra edator-prey model (4.2) studied in Section 4.1.

Note that $D(0) = -b_1 b_2 < 0$. If z is taken to be a positive real number, y $z = x > 0$, then $D(x) = x^2 - b_1 b_2 k_1^*(x) k_2^*(x)$ where the Laplace transforms (x) are bounded for all $x \geq 0$: $|k_i^*(x)| \leq 1$. It follows easily that

$D(+\infty) = +\infty$ and as a result we see that the characteristic equation $D(z) = 0$ has at least one positive real root. This proves the following theorem. (Also see Theorems 4.16 and 4.18.)

THEOREM 4.9 *If* $k_i(t) \varepsilon L_+^1$, $|k_i|_1 = 1$ *under the above assumptions on* f_i, *then the equilibrium* $N_i \equiv e_i$ *of* (4.21) *is unstable.*

As usual this theorem is still valid if one or both $k_i \equiv \delta_0$ in which case $k_i^* \equiv 1$ in the above characteristic function.

This result, as far as it goes, for the case of competing species $b_i > 0$ is commensurate with the well-known "principle of competitive exclusion" and seems to indicate that time delays do not appear to interfere with this principle at least as far as the instability of the equilibrium is concerned.

The model (4.19) with $b_i < 0$ is never used for even an elementary discussion of mutualism since it leads either to total extinction ($N_i \to 0$ as $t \to +\infty$) or unbounded populations ($N_i \to +\infty$ as $t \to +\infty$). As is pointed out by May (1974) mutualistic interactions quite often characteristically involve significant time delays. For example, the effect on a plant species of contacts with a pollinator would involve a delay equal to at least that for which it takes the plant to produce a new generation. Thus one might hope that such delays would yield a stable equilibrium for a model of the form (4.21), a hope which runs contrary to the usual tenet that delays are a destabilizing influence. We see from Theorem 4.9 however that in fact this hope cannot be fulfilled.

We close this section with a brief consideration of the competition model (4.23), $b_i > 0$ when resource limitation terms are taken into account:

$$(4.25) \qquad N_i'/N_i = b_i(1 - c_{ii}\int_{-\infty}^{t} N_i(s)k_{ii}(t-s)ds - c_{ij}\int_{-\infty}^{t} N_j(s)k_{ij}(t-s)ds)$$

$$1 \le i,j \le 2, \quad i \ne j, \quad b_i > 0, \quad c_{ij} > 0.$$

: c_{ii} small compared to c_{ij}, $i \ne j$ we expect instability as found above :h $c_{ii} = 0$. As we will see this turns out to be true. This amounts to saying ıt individuals compete more with those of the other species rather than with ›se of the same species or, in other words, inter-species competition is more ;nificant than intra-species competition. Such an assumption and result is ısistent with the principle of competitive exclusion as derived from the nonde- ' version of this model (see J. M. Smith (1974, p. 59)).

In the opposite case when intra-species competition is the more significant en we would expect stability as is the case for the nondelay version (again see M. Smith (1974)). This will be true when delays are present provided of course :h species are stable in the absence of the other.

The characteristic equation for the equilibrium

$$\text{.26)} \qquad e_1 = (c_{22} - c_{12})/\Delta, \quad e_2 = (c_{11} - c_{21})/\Delta, \quad \Delta = c_{11}c_{22} - c_{12}c_{21}$$

$$D(z) := (z + e_1 b_1 c_{11} k^*_{11}(z))(z + e_2 b_2 c_{22} k^*_{22}(z)) - k(z) = 0$$

$$k(z) := e_1 e_2 b_1 b_2 c_{12} c_{21} k^*_{12}(z) k^*_{21}(z).$$

If c_{ii} are both small then $D(z)$ is nearly $z^2 - k(z)$, the characteristic ıction of (4.23). Thus, using Rouche's Theorem it is again easy to show that :) has roots in the right half plane (under the added condition that $tk_{ij} \varepsilon L^1$).

For c_{ij}, $i \ne j$, both small the equilibrium (4.26) satisfies $= c^{-1}_{ii} + O(c_{ij})$ and $D(z)$ is nearly $p(z) := (z + b_1 k^*_{11}(z))(z + b_2 k^*_{22}(z))$. We ıld like to argue that if $p(z)$ has no roots $\text{Re } z \ge 0$ then neither does $D(z)$

provided only that c_{ij}, $i \neq j$, are both small. Suppose that to the contrary there exist sequences $c_{ij}^{(n)}$, z_n such that

$$c_{ij}^{(n)} \to 0 \quad \text{as} \quad n \to \infty, \qquad \mathrm{Re}\, z_n \geq 0, \qquad D(z_n) = 0.$$

Now $D(z_n) = 0$ implies that z_n is bounded and, hence, without loss in generality we may assume that $z_n \to z_0$ for some z_0, $\mathrm{Re}\, z_0 \geq 0$. By continuity $p(z_0) = 0$, which contradicts our assumption that $p(z) \neq 0$ for $\mathrm{Re}\, z \geq 0$.

On the other hand if $p(z) = 0$ has at least one root $\mathrm{Re}\, z > 0$ (i.e. if at least one of the species is unstable in the absence of the other) then another simple argument using Rouche's Theorem (as in the proof of Theorem 4.6 (i)) shows that $D(z) = 0$ has a root $\mathrm{Re}\, z > 0$.

THEOREM 4.10 _Assume_ $k_{ij} \varepsilon L_+^1$, $|k_{ij}|_1 = 1$ _and_ $tk_{ij}(t) \varepsilon L^1$. _For_ c_{ii} _both small the equilibrium_ (4.26) _of the delay competition model_ (4.25) _is unstable. For_ c_{ij}, $i \neq j$ _both small this equilibrium is_ (_locally_) _A.S. if both species have a_ (_locally_) _A.S. equilibrium in the absence of the other and is unstable if at least one species is unstable in the absence of the other_.

This theorem is clearly also valid for general competition models $N_i'/N_i = b_i f_i(N_1, N_2)$ which have the same linearization as (4.25).

Finally we consider some examples which illustrate a few points regarding time delays in these simple competition models. These illustrations deal with the model (4.25).

(i) _Delays in a competition model_ (4.25) _cannot stabilize an otherwise unstable equilibrium_. Assume that the equilibrium (4.26) is positive. It is unstable for the nondelay version of (4.25) (i.e. when $k_{ij} = \delta_0$) if $c_{11} < c_{21}$ and $c_{22} < c_{12}$ which implies that

27) $$\Delta := c_{11}c_{22} - c_{12}c_{21} < 0.$$

D(0) $= e_1e_2b_1b_2\Delta < 0$ and $D(x) \to +\infty$ as $x \to +\infty$, x real. Thus, $D(z)$ ays has at least one positive, real root. (See Theorem 4.18.)

(ii) Two competing species may possess a stable equilibrium even when one of species in the absence of the other has an unstable equilibrium. To illus-te this point suppose that the only delay present in the model (4.25) is in first species' response to resource limitation: $k_{12}(t) = k_{21}(t) = k_{22}(t) = \delta_0$ $k_{11}(t) = T^{-2}t \exp(-t/T)$, $T > 0$. Assume that intra-species competition is onger than inter-species competition in the sense that

$$c_{22} > c_{12}, \quad c_{11} > c_{21}.$$

s, in order that the equilibrium (4.26) be positive we assume

$$\Delta = c_{11}c_{22} - c_{12}c_{21} > 0.$$

The characteristic function becomes

$$D(z) = \left(z + \frac{\alpha_1}{(zT + 1)^2}\right)(z + \alpha_2) - \beta$$

$$\alpha_1 = e_1b_1c_{11} > 0, \qquad \beta = e_1e_2b_1b_2c_{12}c_{21} > 0.$$

s, $D(z) = N(z)/(zT + 1)^2$ where

$$N(z) := T^2z^4 + (\alpha_2T^2 + 2T)z^3 + (1 + 2\alpha_2T)z^2 + (\alpha_1 + \alpha_2)z + (\alpha_1\alpha_2 - \beta).$$

$\alpha_1\alpha_2 - \beta = e_1e_2b_1b_2\Delta > 0$ so that all coefficients of $N(z)$ are positive.

The equilibrium will be stable when the four Hurwitzian determinants of $N(z)$ are positive. These determinants are

$$H_1 = 2T + \alpha_2 T^2 > 0$$

$$H_2 = 2\alpha_2^2 T^3 + (4\alpha_2 - \alpha_1)T^2 + 2T$$

$$H_3 = (\alpha_1 + \alpha_2)H_2 - (\alpha_1\alpha_2 - \beta)T^2(2 + \alpha_2 T)^2$$

$$H_4 = (\alpha_1\alpha_2 - \beta)H_3.$$

Suppose we consider the case when b_1 is small. As $b_1 \downarrow 0$ we find that $\alpha_1 \downarrow 0$ and $\beta \downarrow 0$ and hence $H_2 \to 2\alpha_2^2 T^3 + 4\alpha_2 T^2 + 2T > 0$ and $H_3 \to \alpha_2 H_2 > 0$.

Thus, the equilibrium is A.S. if c_{11} and c_{22} are large and if $b_1 > 0$ is sufficiently small. However, if $T > 2/\alpha_1$ then N_1 in the absence of N_2 has an unstable equilibrium because of the large delay (cf. Section 3.1).

(iii) Time delays (no matter how small) can reverse the outcome of a competition between two species. This point is made by Caswell (1972) by means of numerical simulations of a more complicated model (involving three species) than (4.25), although it is not clear that it is actually the time delays that cause the reversal in Caswell's work. We will illustrate this point by means of a simple example using the simplest model (4.19) and its delay version (4.23) with, for simplicity, $c_{ij} = 1$.

From the direction field of (4.19) (see FIGURE 4.3) we see that any solution initially satisfying $N_1(0) > 1$, $N_2(0) < 1$ must satisfy $N_1(+\infty) = +\infty$, $N_2(+\infty) = 0$, i.e. N_2 goes extinct and N_1 "wins" the competition.

Consider the delay model (4.23) with $k_2 = \delta_0$ and $k_1(t) = T^{-1} \exp(-t/T)$, $T > 0$ which represents a competition in which the response of one species (namely

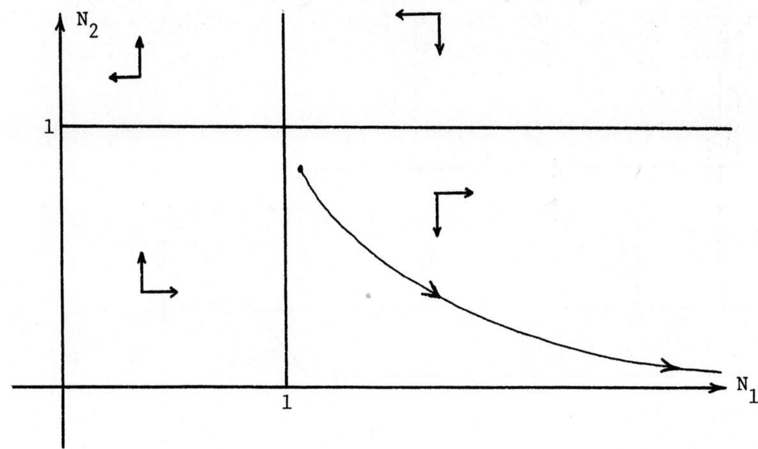

FIGURE 4.3

) is delayed, but only "weakly" delayed.

If we let $Q(t) := \int_{-\infty}^{t} N_1(s)k_1(t - s)ds$ then it is easy to see that any lution of (4.23) with these kernels must satisfy the (nondelay) system

$$N_1' = b_1N_1(1 - N_2), \quad N_2' = b_2N_2(1 - Q), \quad Q' = T^{-1}(N_1 - Q) \tag{.28}$$

rt of whose direction field is indicated in FIGURE 4.4.

Note that if $N_1(0) = N_2(0) = 1$, $Q(0) = 0$ then the trajectory of (4.28) st satisfy $N_1(+\infty) \to 0$, $N_2(+\infty) = +\infty$, $Q(+\infty) = 0$ as drawn in FIGURE 4.4. For y other trajectories starting near this initial point $P(1,1,0)$ the same must ppen. In particular, unlike the nondelay model (4.19), there are trajectories r which $N_1(0) > 1$, $N_2(0) < 1$ $(Q(0) \approx 0)$ for which $N_1(+\infty) = 0$, $N_2(+\infty) = +\infty$ d hence N_2 "wins." This is all independent of T and hence this reversal n occur for arbitrarily small time delays.

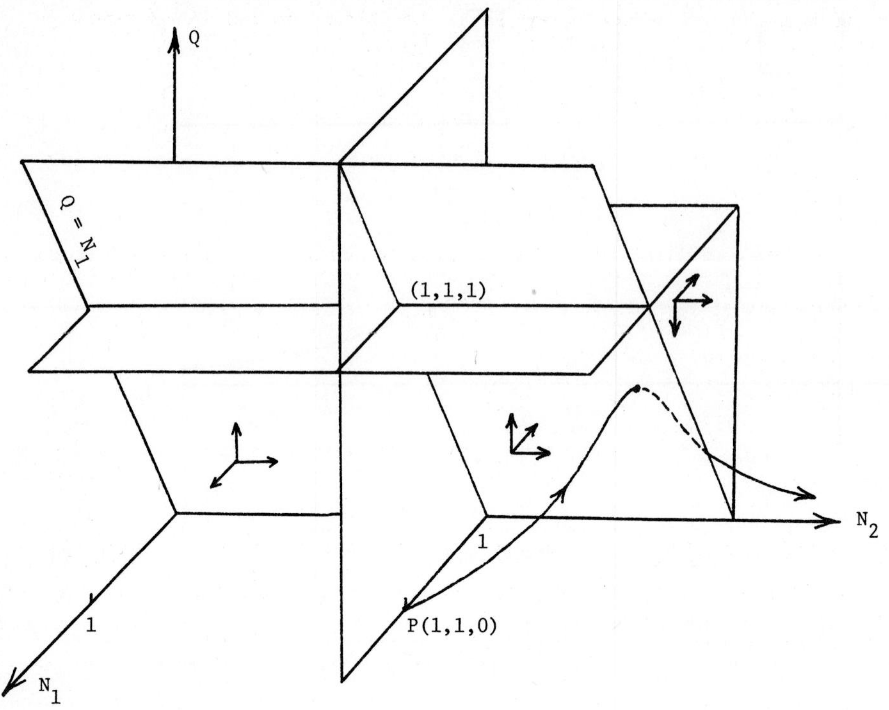

FIGURE 4.4

4.9 Stability and Instability for n-Species Models. It is of course likely that a given species interacts significantly with more than one other species in a community. So far we have only considered the interaction of two species with each other. The addition of a third species introduces a new level to the study of population dynamics, not only in terms of the number of experimental and observable parameters which must be dealt with (Poole (1974)) but also in terms of the complexity of the models and the variety of possible resulting dynamical behavior (Smale (1976)). Because of this we will restrict our attention to a few general results concerning n-species models. (In later sections and chapters we will occasionally consider specific 3-species models of certain specific types.)

Here we consider the general model (1.1), which for convenience we renumber

.29) $$N_i'/N_i = f_i(N_1, \ldots, N_n), \quad 1 \leq i \leq n$$

ere we assume that a positive equilibrium exists: $f_i(e_1, \ldots, e_n) = 0$ for some sitive constants $e_i > 0$ and that when (4.29) is linearized about this equilib-
um we obtain a system of the form

.30) $$x' = \int_0^t K(t - s)x(s)ds$$

r $x = \text{col}\ (x_i)$, $x_i = N_i - e_i$ and for an $n \times n$ matrix kernel
:) $= (a_{ij}k_{ij}(t))$, $k_{ij} \varepsilon L_+^1$, $|k_{ij}|_1 = 1$. (Here a_{ij} is essentially
$\partial f_i(e_1, \ldots, e_n)/\partial N_j$.) The characteristic equation for (4.30) is

.31) $$D(z) := \det\ (zI - K^*(z)) = 0.$$

First suppose we assume that each k_{ij} is a generic, exponential kernel or
re generally that

.32) $$k_{ij}(t) = \sum_{m \geq 0} c_{ij,m} \frac{t^m}{m!T_{ij,m}^{m+1}} \exp\ (-t/T_{ij,m}), \quad T_{ij,m} > 0$$

ere the coefficients $c_{ij,m}$ are finite in number and are such that $k_{ij} \geq 0$
d $\Sigma_m c_{ij,m} = 1$ for all i,j (so that $|k_{ij}|_1 = 1$). Then

$$k_{ij}^*(z) = \sum_{m \geq 0} \frac{c_{ij,m}}{(zT_{ij,m} + 1)^{m+1}}$$

a rational function of z, which has the property that

(4.33) $k^*_{ij}(z) \to 1$ uniformly in z, $\text{Re } z \geq 0$ as all $T_{ij,m}$ tend to zero.

Now

$$D(z) = z^n + d_1 z^{n-1} + \ldots + d_n, \quad d_i = d_i(T_{ij,m};z)$$

and if $p(z) := \det(zI - K_0)$, $K_0 = (a_{ij})$ then

$$p(z) = z^n + d_1^0 z^{n-1} + \ldots + d_n^0, \quad d_n^0 = \text{constant}.$$

It follows from (4.33) that $d_k(T_{ij,m};z) \to d_k^0$ uniformly in z, $\text{Re } z \geq 0$ as all $T_{ij,m}$ tend to zero.

Suppose that $p(z)$ is a stable polynomial. Then by the Hurwitz criteria all $d_k^0 > 0$ and all Hurwitzian determinants $H_k^0 > 0$. It follows that all $d_k > 0$ and $H_k > 0$ (where $H_k = H_k(T_{ij,m};z)$ are the Hurwitzian determinants for d_k) for all z, $\text{Re } z \geq 0$ and all $T_{ij,m} > 0$ sufficiently small, say $|T_{ij,m}| \leq T_0$, $T_0 > 0$. If $D(z)$ had a root $z = z_0$, $\text{Re } z_0 \geq 0$ then so would the polynomial $z^n + d_1(T_{ij,m};z_0)z^{n-1} + \ldots + d_n(T_{ij,m};z_0)$ which is impossible if $|T_{ij,m}| \leq T_0$ since in this case all of the Hurwitzian criteria hold. We have proved the first part of the following theorem.

THEOREM 4.11 (i) *Together with the assumptions made above on the model* (4.29), *assume that all eigenvalues of* $K_0 = (a_{ij})$ *lie in the left half plane. Suppose further that every delay kernel has the general generic form* (4.32). *Then there exists a constant* $T_0 > 0$ *such that* $|T_{ij,m}| \leq T_0$ *for all* i,j,m *implies that the equilibrium of* (4.29) *is (locally) A.S.*

(ii) *If on the other hand* K_0 *has an eigenvalue in the right half plane* $\text{Re } z > 0$, *then there is such a constant* T_0 *such that the equilibrium of* (4.29) unstable.

Proof. We have only to prove (ii). Let $R > 0$ be so large that all roots p(z), which lie in the right half plane, lie in the interior of sector: $z_0 \geq x_0$, $|z| \leq R$ whose boundary we denote by $\partial(R)$ where $x_0 > 0$ is chosen that $p(x_0 + iy) \neq 0$ for all y. Let $m = \min_{\partial(R)} |p(z)|$, which is a posi-e number $m > 0$. Using (4.33), which implies $- d_k^0| \leq m/2(R^{n-1} + R^{n-2} + \ldots + 1)$ for $z \varepsilon \partial(R)$ and $|T_{ij,m}| \leq T_0$ for some ll $T_0 > 0$, we find that

$$D(z) - p(z)| \leq |d_1 - d_1^0||z|^{n-1} + \ldots + |d_n - d_n^0| \leq m/2 < |p(z)|, \quad z \varepsilon \partial(R).$$

che's Theorem implies D(z) has roots inside the sector and hence in the right f plane. □

The matrix K_0 can be thought of as the coefficient matrix of the linearized sion of (4.29) when delays are absent $(k_{ij} = \delta_0)$. Thus, Theorem 4.11 says ghly that the stability or instability of the equilibrium of the general model h small delays is that of the equilibrium when delays are absent. Here the $_{,m}$ measure the "length" of the delay in the kernels (4.32).

Next we generalize some of the arguments we used above to prove theorems ut predator-prey models to the model (4.29) (or more precisely to the linear-tion (4.30)). First we use the Argument Principle on the characteristic equa-n (4.31) in order to derive general, geometric stability and instability teria for n-species interaction models with delayed growth rate responses.

The characteristic function of the linearization (4.30) is given by $) = z^n + d_1 z^{n-1} + \ldots + d_n$ where $d_k = d_k(z)$ is bounded for $\operatorname{Re} z \geq 0$ (since $k_{ij}^*(z)$ are). Suppose $tk_{ij}(t) \varepsilon L^1$. Then $(d/dz)k_{ij}^*(z)$ is also bounded for

$\mathrm{Re}\, z \geq 0$. Differentiation yields

$$\frac{D'(z)}{D(z)} = \frac{(n + d_1')z^{n-1} + ((n-1)d_1 + d_2')z^{n-2} + \ldots + (2d_{n-2} + d_{n-1}')z + (d_{n-1} + d_n')}{z^n + d_1 z^{n-1} + \ldots + d_n}$$

and

$$\frac{D'(z)}{D(z)} - \frac{n}{z} = \frac{d_1' z^n}{zD(z)} + \frac{(d_2' - d_1)z^{n-1} + \ldots - nd_n}{zD(z)} := R_1(z) + R_2(z).$$

Suppose $\partial^1(R) = \{z\colon z = R\exp(i\theta),\ -\pi/2 \leq \theta \leq \pi/2\}$. It is easy to see that

$$|R_2(z)| \leq M \frac{R^{n-1} + \ldots + 1}{R^{n+1} - M(R^n + \ldots + 1)}$$

for some constant $M > 0$ so that

$$\left|\int_{\partial^1(R)} R_2(z)dz\right| \leq M \frac{R^{n-1} + \ldots + 1}{R^{n+1} - M(R^n + \ldots + 1)} \pi R \to 0 \quad \text{as} \quad R \to +\infty.$$

Also

$$\left|\int_{\partial^1(R)} R_1(z)dz\right| \leq \frac{R^{n+1}}{R^{n+1} - M(R^n + \ldots + 1)} \int_{-\pi/2}^{\pi/2} |d_1'(R\exp(i\theta))|d\theta \to 0$$

as $R \to +\infty$ provided

(4.34) $\quad d_1'(R\exp(i\theta)) \to 0$ as $R \to +\infty$ uniformly in $-\pi/2 \leq \theta \leq \pi/2$.

As a result we deduce that

$$\lim_{R \to +\infty} \int_{\partial^1(R)} \frac{D'(z)}{D(z)} dz = \lim_{R \to +\infty} \int_{\partial^1(R)} \frac{n}{z} dz = n\pi i.$$

Let $\partial(R)$ be the boundary of the semi-circle $\{z: |z| \leq R, \ \mathrm{Re}\ z \geq 0\}$. The
;ument Principle implies, provided

$$D(z) \neq 0 \quad \text{for} \quad \mathrm{Re}\ z = 0, \tag{35)}$$

ıt the number of roots of $D(z)$ with $\mathrm{Re}\ z \geq 0$ is

$$\nu(+\infty) = \lim_{R \to +\infty} \frac{1}{2\pi i} \int_{\partial(R)} \frac{D'(z)}{D(z)}\, dz$$

$$= \frac{1}{2\pi} \lim_{R \to +\infty} (\arg D(-iR) - \arg D(iR)) + \frac{n}{2}$$

ɛre we have taken the principle branch of the log function $-\pi < \arg z \leq \pi$
ler the assumption that

$$D(0) = (-1)^n \det K_0 > 0, \quad K^0 = (a_{ij}) \tag{36)}$$

that $D(0)$ does not lie on the cut along the negative real axis. (If
$)^n \det K_0 < 0$ then the equilibrium is unstable as will be pointed out in
ıark 2 below.) Since $k_{ij}(\bar{z}) = \overline{k_{ij}(z)}$ it easily follows that $D(\bar{z}) = \overline{D(z)}$ so
ıt $\arg D(-iR) = -\arg D(iR)$ and we get

$$\nu(+\infty) = \frac{n}{2} - \frac{\arg D(+i\infty)}{\pi} .$$

ıs leads us to the following generalization of Theorems 4.1 and 4.8.

THEOREM 4.12 <u>Consider the general model</u> (4.29) <u>under the conditions described</u>
ɔve. <u>Suppose that the delay kernels satisfy</u> $k_{ij} \varepsilon L^1_+$, $|k_{ij}|_1 = 1$, $tk_{ij}(t) \varepsilon L^1$
l <u>that</u> (4.34), (4.35) <u>and</u> (4.36) <u>hold</u>. <u>Then the equilibrium is</u> (<u>locally</u>) <u>A.S.</u> <u>if</u>

$\arg D(+i\infty) = n\pi/2$ _and unstable if_ $\arg D(+i\infty) \neq n\pi/2$.

Remark (1) The highest order term (in R) in $D(iR)$ is $i^n R^n$ (again we use the boundedness of $k^*_{ij}(z)$ for $\text{Re } z \geq 0$). Thus we can distinguish two cases depending on the number of species n:

$$\arg D(+i\infty) = \begin{cases} \frac{2k+1}{2}\pi & \text{for } n \text{ odd and some } k \leq (n-1)/2 \\ k\pi & \text{for } n \text{ even and some } k \leq n/2. \end{cases}$$

We conclude the following corollary.

COROLLARY 4.13 _Under the hypotheses of Theorem_ 4.12 _the equilibrium of the general model_ (4.29) _is_ (_locally_) _A.S. if and only if_ $k = (n-1)/2$ _for_ n _odd or_ $k = n/2$ _for_ n _even where_ n _is the number of species_.

Remark (2) With regard to the condition (4.36) we observe that for real $z = x > 0$, $D(x) = O(x^n)$ for large x and hence $D(+\infty) = +\infty$. If $D(0) < 0$ then D has a positive real root.

THEOREM 4.13 _If_ $k_{ij} \varepsilon L^1_+$, $|k_{ij}|_1 = 1$ _and_ $(-1)^n \det(a_{ij}) < 0$ _then the equilibrium of_ (4.29) _is unstable_.

The Theorems 4.12 and 4.13 remain valid if any or all $k_{ij} = \delta_0$, that is when no delays are present in some interactions in the linearization.

Remark (3) Note that the coefficient d_1 in $D(z)$ is the trace of $K^*(z)$: $d_1 = a_{11}k^*_{11}(z) + \ldots + a_{nn}k^*_{nn}(z)$. Thus, condition (4.34) is a condition involving only the self-damping or resource limitation terms in the linearized model. This condition is clearly satisfied if either no self-damping is present

1 $a_{ii} = 0$) or if any self-damping in the model is nondelayed ($k^*_{ii}(z) = 1$ so t d_i constant). It is also easy to see that (4.34) is satisfied if the ker- s k_{ii} are of the general generic form (4.32). □

As in Sections 3.4 and 4.7 for single species and predator-prey models we study the general model (4.29) as a function of some measure of the delay in system in relation to the inherent growth rates of the species (May et al. 74b), May (1973)). Thus we consider the system

37) $$N_i'/N_i = b_i f_i(N_1, \ldots, N_n), \quad 1 \leq i \leq n$$

er the same general assumptions on f_i made above where $b_i \neq 0$ is to be ught of as the inherent unrestrained growth rate of the ith species. Let 0 be some measure of the delay in (4.37) (such as the average of all $T_{ij,m}$ kernels of the form (4.32)) and make the time scale change $t^* = t/T$ in (4.37) ch then becomes

38) $$N_i'/N_i = b_i T f_i(N_1, \ldots, N_n), \quad 1 \leq i \leq n$$

ch has the linearization (4.30) with kernel $K(t) = (Tb_i a_{ij} k_{ij}(t))$. We do this to enter explicitly the time delay T into the analysis by choosing it as unit of time.

Let m be a fixed integer $1 \leq m \leq n$ and let $K_m(t)$ denote the submatrix K(t) obtained by deleting the mth row and column from $K(t)$. Let z) = $\det(zI_{n-1} - K^*_m(z))$ where I_{n-1} is the $n - 1 \times n - 1$ identity. Fix for $i \neq m$. Define $u = b_m T$ and consider $D = D(z,u)$ as a function of z u.

Note that $D(0,0) = 0$. Using the familiar "row-by-row" differentiation formula for determinants we find that

$$D_z(0,0) = (-1)^{n-1} \det K_m^*(0) = (-T)^{n-1}\left(\prod_{j\neq m} b_j\right) \det A_m$$

where A_m is the submatrix of $A = (a_{ij})$ obtained by deleting the mth row and column. Also we get

$$D_u(0,0) = -(-T)^{n-1}\left(\prod_{j\neq m} b_j\right) \det A$$

so that if

$$\text{(4.39)} \qquad \det A \neq 0 \quad \text{and} \quad \det A_m \neq 0 \quad \text{for some} \quad m, \quad 1 \leq m \leq n$$

then by the implicit function theorem there exists a unique solution branch of the characteristic equation $D(z,u) = 0$

$$\text{(4.40)} \qquad z = z(u), \quad z(0) = 0, \quad |u| \leq u_0, \quad u_0 > 0, \quad z'(0) = r_m \neq 0$$

$$r_m := \det A/\det A_m .$$

THEOREM 4.14 _In addition to the assumptions on_ f_i _in the general delay model_ (4.38) _made above, assume that the delay kernels satisfy_ $k_{ij} \varepsilon L_+^1$, $|k_{ij}|_1 = 1$, $tk_{ij}(t)\varepsilon L^1$ _and that_ (4.39) _holds for some_ m. _Given values of_ b_iT _for_ $i \neq m$ _there exists a constant_ $u_0 > 0$ _such that for_ $|b_mT| \leq u_0$ _the equilibrium of_ (4.38) _is_ (i) _unstable when_ $\operatorname{sign} b_m = \operatorname{sign} r_m$ _and_ (ii) _is, if in addition_ $p_m(z) \neq 0$ _for_ $\operatorname{Re} z \geq 0$, _locally A.S. when_ $\operatorname{sign} b_m = -\operatorname{sign} r_m$. Here $\operatorname{sign} b_m := b_m/|b_m|$.

Proof. (i) From (4.40) follows $\operatorname{Re} z(u) > 0$ for $\operatorname{sign} u = \operatorname{sign} r_m$,

$| \leq u_0$, for u_0 smaller if necessary.

(ii) From (4.40) follows $\text{Re } z(u) < 0$ for $\text{sign } u \neq -\text{sign } r_m$ and u all. Suppose, for purposes of contradiction, that D has at least one root $z \geq 0$ for arbitrarily small u. Then there exists sequences $u_n \to 0$, z_n th $\text{Re } z_n \geq 0$, $D(z_n,u_n) = 0$. Since z_n unbounded implies $D(z_n,u_n)$ is un-unded it follows that z_n is a bounded sequence and that we may assume (by tracting a subsequence if necessary) that $z_n \to z_0$, $\text{Re } z_0 \geq 0$. By continuity $z_0,0) = 0$. But $D(z_0,0) = z_0 p_m(z_0) = 0$ implies $z_0 = 0$. The existence of such sequence $u_n \to 0$, $z_n \to 0$, $\text{Re } z_n \geq 0$ contradicts the uniqueness of the solu-on branch (4.40). □

The purpose of Theorem 4.14 above is to study the stability (or instability) the equilibrium of a general multi-species model (4.38) as it is a function of e relationship between the delay in the system's responses and the inherent owth rate b_m of one of the member species. The hypotheses in (4.39) amount requiring that both the n-species community and the $(n - 1)$-species subcom-ity obtained by eliminating the mth species both have nonsingular community trices (and hence have isolated equilibria). The crucial condition in Theorem 14 is on the sign of the inherent growth rate b_m, that is to say is whether e m^{th} species grows or dies exponentially in the absence of all inter- and ra-species interactions. The condition that $p_m(z) \neq 0$ for $\text{Re } z \geq 0$ means at the isolated equilibrium of the $(n - 1)$-subcommunity is A.S. (This, inci-tally implies that $0 < p_m(0) = (-1)^{n-1}K_m^*(0)$ or $\text{sign det } A_m = (-1)^{n-1}$.)

This approach could be extended in a rather obvious manner to the case when or more b_iT are small. We will not do this here since a more generalized sion of Theorem 4.14 than this will be given in Theorem 4.16 below.

It is also possible to generalize Theorem 3.1 and its proof to the more neral case $n > 1$ of (3.29). This approach deals with the magnitude of the response to interactions with delays compared to that of those without delays, instead of with the "length" of delay as was the intent in Theorem 4.14 above. Suppose the linearization of (4.29) has the form

$$x' = K_0 x + \int_0^t \bar{K}(t - s)x(s)ds \tag{4.41}$$

instead of (4.30) where K_0 is a constant matrix and $\bar{K} = (a_{ij}\bar{k}_{ij}(t))$, $\bar{k}_{ij} \varepsilon L_+^1$, $|\bar{k}_{ij}|_1 = 1$. In this model the instantaneous response K_0 is separated from the delayed response $\bar{K}$. The characteristic function can be written

$$D(z) := \det(zI - K_0 - \bar{K}^*(z)) = z^n + d_1 z^{n-1} + \ldots + d_n$$

where $d_m \to d_m^0$ as every $a_{ij} \to 0$ uniformly for $\text{Re } z \geq 0$ where d_m^0 is the coefficient of z^{n-m} in the polynomial $p(z) := \det(zI - K_0)$. Arguing just as in the proof of Theorem 4.11 (with a_{ij} in place of $T_{ij,m}$) we obtain

THEOREM 4.15 <u>In addition to the assumptions made on</u> f_i <u>above (where now</u> (4.40) <u>is the linearization) assume</u> K_0 <u>has no purely imaginary eigenvalues. There exists a constant</u> $a_0 > 0$ <u>such that</u> $|a_{ij}| \leq a_0$, $1 \leq i,j \leq n$ <u>implies that the equilibrium of</u> (4.29) <u>is</u> (i) <u>unstable if</u> K_0 <u>has eigenvalues in the right half plane and</u> (ii) <u>(locally) A.S. if all eigenvalues are in the left half plane</u>.

Thus, the asymptotic stability or instability of the delay model is the same as that of the nondelay version of the model when the magnitude of the delayed growth rate response is small.

All of the above theorems (except Theorem 4.12) deal with the case when the delays or their effects are in some sense small. We expect as delays become more significant that the equilibrium of the general model (4.38) will become unstable.

hat this is not always the case can be seen by Theorem 3.3 for $n = 1$.) We con-
ude this section with a theorem which given conditions under which this is true.

Consider system (4.38) for which the delay measure T is the unit of time.
t $u_i = b_i T$ and $u = col(u_i)$. The characteristic function $D: = \det(zI - K^*(z))$,
t) $= (u_i a_{ij} k_{ij}(t))$ is a function of z and u: $D = D(z,u)$. Assume that

.42) there exists a vector u_0 and a real $y > 0$ such that $D(iy,u_0) = 0$.

ote: $D(\bar{z},u_0) = \overline{D(z,u_0)}$ shows that the restriction $y > 0$ is no loss in gener-
ity.)

This assumption means that for some critical values of the parameters $b_i T$
e equilibrium is "marginally" stable, i.e. the characteristic equation has a
rely imaginary root.

We wish to give conditions under which $D(z,u)$ has roots in the right half
ane for u near the critical value u_0. Let b be a unit vector. Assume

.43) $$D_z(iy,u_0) \neq 0 \quad \text{and} \quad \rho: = \text{Re}[b\cdot\nabla_u D(iy,u_0)/D_z(iy,u_0)] \neq 0$$

ere $w\cdot b$ denotes the usual Euclidean dot or scalar product. The first condi-
on in (4.43) and the implicit function theorem guarantee that $D(z,u) = 0$ can
solved uniquely for $z = z(u)$ for $|u - u_0| \leq u^*$, $u^* > 0$ small, such that
$u_0) = iy$. Implicit differentiation yields

$$b\cdot\nabla z(u_0) = -b\cdot\nabla_u D(iy,u_0)/D_z(iy,u_0).$$

follows from the second condition in (4.43) that $\text{Re}\, z(u) > 0$ for u close
u_0 in the direction of b (or $-b$) if $\rho < 0$ (or $\rho > 0$).

THEOREM 4.16 *In addition to the assumption made on* f_i *above, assume that the delay kernels satisfy* $k_{ij} \varepsilon L^1_+$, $|k_{ij}|_1 = 1$, $tk_{ij}(t) \varepsilon L^1$ *and that* (4.42) *and* (4.43) *hold. Then the equilibrium of the general delay model* (4.38) *is unstable for* $u = \mathrm{col}(b_i T)$ *close to* u_0 *and in the direction of* b (*or* $-b$) *when* $\rho < 0$ (*or* $\rho > 0$).

The conditions (4.42) and (4.43) in their stated generality are a little difficult to relate directly to the parameters and delay kernels in the (linearized) system. To make this relationship more explicit at least in a special case we consider the case when the delay is significant only when compared to one growt rate, say b_1, while it is small in comparison to the others b_i, $i \neq 1$. Thus w assume $b_i T = \varepsilon \beta_i$, $i \neq 1$, for fixed constants β_i and a small constant $\varepsilon > 0$. We now view D as a function of z, $u_1 = b_1 T$ and ε: $D = D(z,u_1,\varepsilon)$.

First consider the condition (4.42). Now

$$D(iy,u_1,\varepsilon) = (iy)^{n-1}[iy - u_1 a_{11} k^*_{11}(iy)] + 0(\varepsilon).$$

Writing

$$k^*_{11}(iy) = C_{11}(y) - iS_{11}(y), \quad C_{11} = \int_0^\infty k_{11}(t)\cos yt dt, \quad S_{11} = \int_0^\infty k_{11}(t)\sin yt dt$$

we have that the equation $D(iy,u_1,\varepsilon) = 0$ for $y > 0$ is equivalent to the two real equations

$$(4.44) \qquad u_1 a_{11} C_{11}(y) + 0(\varepsilon) = 0, \quad y + u_1 a_{11} S_{11}(y) + 0(\varepsilon) = 0.$$

If we assume that

.45) $a_{11} \neq 0$ and there exists a $y_0 > 0$ such that $C_{11}(y_0) = 0$, $S_{11}(y_0) \neq 0$

en the equations (4.44) are satisfied for $y = y_0$, $u_1 = u_1^0$, $\varepsilon = 0$ where

.46) $$u_1^0 = -y_0/a_{11}S_{11}(y_0).$$

n (4.44) be solved for small $\varepsilon > 0$? The Jacobian of these two equations with spect to u_1, y evaluated at u_1^0, y_0 and $\varepsilon = 0$ turns out to be equal to $a_{11}\int_0^\infty tk_{11}(t)\sin y_0 t dt$ and hence if

.47) $$\sigma_S(y_0) := \int_0^\infty tk_{11}(t)\sin y_0 t dt \neq 0$$

en (4.44) can be solved for $u_1 = u_1(\varepsilon)$, $y = y(\varepsilon)$, for $\varepsilon > 0$ small where $(0) = u_1^0$, $y(0) = y_0$. This means that hypothesis (4.42) of Theorem 4.16 is tisfied for such u_1, y and ε.

In order to apply Theorem 4.16 we have yet to fulfill the second and final pothesis (4.43) with $b = \text{col}(1, 0, \ldots, 0)$ (since we wish to change only $= b_1T$ while keeping u_i, $i \neq 1$ fixed). Now $z, u_1, \varepsilon) = z^{n-1}(z - u_1 a_{11} k_{11}^*(z)) + O(\varepsilon)$ so

$$(iy, u_1, \varepsilon) = (n - 1)(iy)^{n-2}(iy - u_1 a_{11} k_{11}^*(iy)) + (iy)^{n-1}(1 - u_1 a_{11} k_{11}^{*\prime}(iy)) + O(\varepsilon).$$

: $y = y_0$, $u_1 = u_1^0$ and $\varepsilon = 0$ this yields

.48) $$D_z(iy_0, u_1^0, 0) = (iy_0)^{n-1}[1 + u_1^0 a_{11}\sigma_C(y_0) - iu_1^0 a_{11}\sigma_S(y_0)]$$

ere

$$\sigma_C(y_0):=\int_0^\infty tk_{11}(t)\cos y_0 t dt.$$

Secondly,

$$D_{u_1}(iy,u_1,\varepsilon) = -(iy)^{n-1}a_{11}k^*_{11}(iy) + O(\varepsilon)$$

which for $y = y_0$, $u_1 = u_1^0$ and $\varepsilon = 0$ yields $D_{u_1}(iy_0,u_1^0,0) = (iy_0)^{n-1}a_{11}iS_{11}(y$ Let $\rho^0 := \mathrm{Re}\,[D_{u_1}(iy_0,u_1^0,0)/D_z(iy_0,u_1^0,0)]$. Then we have by these calculations th

$$\rho^0 = \frac{y_0 a_{11}\sigma_S(y_0)}{(1+u_1^0 a_{11}\sigma_C(y_0))^2 + (u_1^0 a_{11}\sigma_S(y_0))^2} \neq 0. \tag{4.49}$$

In (4.43), $\rho = \rho^0 + O(\varepsilon)$ so that from (4.47) and (4.49) it follows that hypothes (4.43) of Theorem 4.16 holds if $\varepsilon > 0$ is small.

COROLLARY 4.17 *If, in addition to the hypotheses of Theorem* 4.16 *on* f_i *and* k_{ij}, *the hypotheses* (4.45) *and* (4.47) *hold then the equilibrium of* (4.38) *is unstable for* $u_i = b_iT$ *small,* $i \neq 1$ *and for* $u_1 = b_1T$ *near* u_1^0 *given by* (4.46) *and* $u_1 > u_1^0$ *(or* $< u_1^0$*) if* $a_{11}\sigma_S(y_0) < 0$ *(or respectively* > 0).

This corollary generalizes Theorem 3.6. Note that the hypotheses (4.45) and (4.47) depend only on the delay kernel k_{11} in the self-inhibition term of that species $n = 1$ whose parameter b_1T is "large." Since the conditions (4.45) and (4.47) require that some delay be present in k_{11} (both fail if $k_{11} = \delta_0$) this corollary means very roughly that if "large enough" delay is present in the growth rate response to resource limitation (i.e. intra-species contacts) of one species in a multi-species community and if this delay is "short" compared to the inherent growth rates of the remaining species in the community, then the equilibrium will be unstable.

As an example of delay kernels which satisfy the conditions (4.45) and (4.47) nsider first the "strong" generic delay kernel $k_{11}(t) = t \exp(-t)$. Then for > 0 we have

$$C_{11}(y) = (1 - y^2)/(1 + y^2), \quad S_{11}(y) = 2/(1 + y^2)^2 > 0$$

$$\sigma_S(y) = -2 \frac{3y + y^3}{(1 - 3y^2)^2 + (3y + y^3)^2} < 0$$

that (4.45) and (4.47) hold if and only if $y_0 = 1$ in which case the critical lue of $u_1 = b_1T$ is given by (4.46): $u_1^0 = -2/a_{11}$. If, as is usually assumed, tra-species contacts have only a negative effect on the growth rate, then $_1 < 0$ and $u_1^0 > 0$ as well as $a_{11}\sigma_S(y_0) > 0$. Thus, instability of the multi- ecies community's equilibrium occurs for $b_1T > u_1^0 = -2/a_{11}$ (and b_iT "small" r $i \neq 1$).

Finally, suppose k_{11} is the "weak" generic delay kernel $k_{11}(t) = \exp(-t)$. en

$$C_{11}(y) = 1/(1 + y) \neq 0, \quad y > 0$$

that (4.45) fails to hold. This means that as b_1T is increased, no change stability or instability is seen (as b_iT remain fixed).

4.10 <u>Delays Can Stabilize an Otherwise Unstable Equilibrium</u>. The results of e previous Section 4.9 do not deal with the question of whether an equilibrium ich is unstable in the absence of delays can be stable when delays are present. know from the results in that section that the answer is "no" if the delays are mall" (in one of several senses). In this section we will prove that for $n = 2$ at delays cannot stabilize an otherwise unstable equilibrium (at least if the trace $a_{11} + a_{22}$ of the self-density responses is negative). We will, however, show by means of a specific example that *when* $n \geq 3$ *delays can stabilize an otherwise unstable equilibrium*.

THEOREM 4.18 *Consider the general model* (4.29) *for* $n = 2$ *under the assumptions that a positive equilibrium exists and that the linearization of* (4.29) *at this equilibrium is of the form* (4.30) $k_{ij} \varepsilon L^1_+$, $|k_{ij}|_1 = 1$, $1 \leq i,j \leq 2$. *If* $a_{11} + a_{22} < 0$ *and if the equilibrium is unstable when delays are absent (in at least the linearization):* $k_{ij} = \delta_0$, *then the equilibrium of* (4.29) *is unstable*.

Proof. With $k_{ij} = \delta_0$ we find

$$D(z) = z^2 - (a_{11} + a_{22})z + \Delta, \quad \Delta = a_{11}a_{22} - a_{12}a_{21} \quad \text{with} \quad a_{11} + a_{22} < 0.$$

It is not difficult to see that the quadratic polynomial D has roots in the right half plane if and only if $\Delta < 0$. The result follows from Theorem 4.13. □

As examples, we see that unstable equilibria of two-species competition and mutualism models cannot be made stable by the inclusion of delays. (See Section (4.8.)

Consider the following specific 3-species model

$$\text{(4.50)} \quad \begin{aligned} &\text{(a)} \quad N_1'/N_1 = 4(1 - N_1 - 3N_2) \\ &\text{(b)} \quad N_2'/N_2 = 4(1 - N_2 - 3N_3) \\ &\text{(c)} \quad N_3'/N_3 = 4(1 - 3N_1 - N_3). \end{aligned}$$

e have chosen these particular types of interactions and specific interaction co-fficients and birth rates only in order to illustrate the point that an unstable quilibrium can be stabilized by delays. Thus we arrived at (4.50) purely as a athematical exercise and not with any particular ecological interpretation in ind. Nonetheless we can interpret the model as follows: each species grows ogistically with carrying capacity equal to one in the absence of the others and ach hampers (say by killing) one of the others growth rate (in a cyclic manner: $_2$ hampers N_1, N_3 hampers N_2 and N_1 hampers N_3) when placed in interac-ion, but with no effect on its own growth rate. Be this as it may, we easily ind that (4.50) has a unique positive equilibrium

$$e_1 = e_2 = e_3 = \frac{1}{4} \tag{4.51}$$

nd that the characteristic equation of the linearization about this equilibrium s $(z + 1)^3 + 27 = 0$. The roots of this equation are $z = -4$, $1/2 \pm 3i\sqrt{3}/2$ nd consequently the equilibrium (4.51) is unstable.

Suppose that a delay is placed in one growth rate response, say that equation a) in (4.50) is replaced by

$$\text{(a')} \qquad N_1'/N_1 = \frac{1}{4}\left(1 - N_1 - 3\int_0^\infty N_2(t - s)k(s)ds\right)$$

ith the generic delay kernel

$$k(t) = \frac{1}{T^2}\, t \exp(-t/T), \qquad T > 0.$$

e characteristic equation becomes

$$D(z) := (z + 1)^3 + \frac{27}{(Tz + 1)^2} = \frac{N(z)}{(Tz + 1)^2}$$

where

$$N(z) := T^2z^5 + T(2 + 3T)z^4 + (1 + 6T + 3T^2)z^3$$
$$+ (3 + 6T + T^2)z^2 + (3 + 2T)z + 28.$$

The roots of $D(z)$ are of course those of $N(z)$. If we set $N(z) = 0$, divide both sides by T^2 and let $\varepsilon = 1/T$ we get the equation

$$z^5 + (3 + 2\varepsilon)z^4 + (3 + 6\varepsilon + \varepsilon^2)z^3 + (1 + 6\varepsilon + 3\varepsilon^2)z^2 + (2\varepsilon + 3\varepsilon^2)z + 28\varepsilon^2 = 0 \tag{4.52}$$

whose roots are identical with those of $D(z)$. All the coefficients of (4.52) are positive and the five Hurwitzian determinants are easily found to be of the form $H_1 = 3 + 2\varepsilon$,

$$H_2 = 8 + \ldots, \quad H_3 = 8 + \ldots, \quad H_4 = 16\varepsilon + \ldots, \quad H_5 = 448\varepsilon^3 + \ldots$$

where the dots indicate terms of higher order in ε than those displayed. Thus, all are positive for ε small, i.e. for T large.

We conclude <u>that although the</u> (<u>nondelayed</u>) <u>system</u> (4.50) <u>has an unstable equilibrium</u> (4.51) <u>the delayed version with</u> (a') <u>replacing</u> (a) <u>with the</u> "<u>strong</u>" <u>generic delay kernel has a</u> (<u>locally</u>) <u>A.S. equilibrium for large enough delay</u> $T > 0$.

If, instead, the "weak" generic delay kernel $k(t) = T^{-1}\exp(-t/T)$, $T > 0$ is used in (a'), then we find that $D(z) = N(z)/(Tz + 1)$ where

$$N(z) := Tz^4 + (1 + 3T)z^3 + 3(1 + T)z^2 + (3 + T)z + 28.$$

If we again divide the equation $N(t) = 0$ by T and let $\varepsilon = 1/T$ we find that he Hurwitzian determinants turn out in this case to have the form $\hat{H}_1 = 3 + \varepsilon$,

$$H_2 = 8 + \ldots, \quad H_3 = 8 + \ldots, \quad H_4 = 224\varepsilon + \ldots .$$

hus, the above statement is still true when the "weak" generic delay kernel is sed. This means that it is possible for even a weak delay to stabilize an other- ise unstable equilibrium.

CHAPTER 5. OSCILLATIONS AND SINGLE SPECIES MODELS WITH DELAYS

The first four chapters have dealt with the stability of equilibria of ecological models which incorporate time delays either in the growth rate response to interactions with other species in a community or in the self-inhibitory response of a species to resource limitations. In this chapter (and the next) we will consider a few topics dealing with oscillatory behavior of solutions. Oscillations in an ecological community could arise of course from a great number of conceivable causes, including such things as the form of the growth rate response itself as it depends on the density of the species; variations in the system's parameters (e.g. birth and death rates, carrying capacities, etc.) due to periodic fluctuations in the environment caused by, for example, seasonal or daily changes in temperature, rain fall, etc.; seasonal or periodic harvesting, seeding, immigration or emigration; random fluctuations of the model parameters; and many other causes. Another source of oscillations can also be delays present in growth rate response to changes in species densities.

One of the predominant themes in the previous chapters was that if delays in growth rate responses are in some sense "small" then the equilibrium of a model will have the stability or instability of that of the same model without delays. This of course is reasonable. By "small delays" we have meant a variety of things: that the magnitude of the responses with delays is small compared to those responses which are essentially instantaneous; that the length of time for changes in species densities to have their (maximal) effect is small when compared to other time scales (specifically, those of the inherent unrestrained growth rates); or that the distribution or weighting of delayed responses be monotonically decreasing into the past.

In this chapter we want to consider the case of an isolated single species whose growth rate response to changes in its own density has a "significant" delay and to see how this leads to oscillations in species density. In Section 5.1 we will study the structure of solutions of general single species models on a short time interval when the delay is significant in comparison to the inherent unrestrained growth rate. In Sections 5.2 and 5.3 the existence (in the presence of an equilibrium) of sustained oscillations (periodic solutions) due to the presence of delays will be considered. Finally in Section 5.4 periodic oscillations due to periodic fluctuation of the environmental parameters is briefly discussed for general delay models.

These results together with the stability results of Chapter 3 show how time delays may possibly help to explain some of the observations often made in laboratory experiments. A common experiment is that of isolating a population of a single species in a favorable laboratory environment with a constant, but limited source of food. Typically one of a variety of different outcomes are observed (May et al. (1974)): the population may monotonically increase to a constant saturation level and persist (as in the case of "logistic growth") or it may oscillate, sometimes persistently and sometimes wildly, often causing ultimate extinction.

As we have just pointed out, the results of this chapter and of Chapter 3 demonstrate how delays in growth rate responses to population density changes and their relationship to other model parameters can be a possible cause of any of these various types of qualitative behavior, even in the simplest of one species growth models.

5.1 Single Species Models and Large Delays. As the "length" of the delay T in growth rate response to species density is increased in a general model

$$N'/N = bf(N)(t), \quad b > 0$$

with an equilibrium $N \equiv e > 0$ which is A.S. for small delays, we expect eventually to see the onset of instability (cf. Sections 3.4 and 3.5) or at least more and more pronounced oscillations. Our purpose in this section is to study the nature of these oscillations at least for a small finite interval of time and for large values of the delay T. We will develop expressions for solutions (starting near equilibrium) which are valid for short time intervals. These expressions, which roughly speaking are expansions in powers of $1/bT$, will to the lowest order show divergent oscillations around the equilibrium e. This is not to say that the equilibrium is necessarily unstable or that sustained oscillations (periodic solutions) do not exist. The expansions will be valid only for short time intervals and hence do not yield information about the nature of solutions for large time values t. Numerically computed solutions described briefly below show the extent to which the first order, oscillatory divergent approximations serve as accurate approximations to the solutions, at least in the example considered and at least for short time intervals. Of course, if the oscillations of the solution around the equilibrium e are strongly divergent, swinging periodically closer and closer to $N = 0$, then it would be only a finite time interval on which the model is valid or alternatively the model would then predict extinction in a short time. In this case it would be reasonable to study the solutions, as we do here, only on short time intervals.

We wish to study the behavior of solutions as a function of the relationship between the two time scales $1/b$ and T. Thus, as in Sections 3.1, 3.4 and 3.5, we suppose that some reasonable measure $T > 0$ of the delay in the functional equation above has been made and then used as a unit of time. This results in an equation of the same form as above with b replaced by bT.

We consider the general model

$$(5.1) \qquad N'/N = bTf\left(\int_{-\infty}^{t} N(s)k(t-s)ds\right), \quad b > 0$$

under the assumption that $k \varepsilon L^1_+$, $|k|_1 = 1$ and

.2) $\left\{ \begin{array}{l} f(\cdot): R^+ \to R \text{ is twice continuously differentiable in some} \\ \text{neighborhood of a positive zero } e > 0: f(e) = 0 \text{ and satis-} \\ \text{fies } f'(e) < 0. \end{array} \right.$

Mathematically our approach will be to use some simpler ideas from singular rturbation theory. And although we confine our attention to (5.1) the method certainly applicable to other or even more general models.

If both sides of (5.1) are divided by the dimensionless parameter bT we get

.3) $$\varepsilon N'/N = f\left(\int_{-\infty}^{t} N(s)k(t-s)ds \right)$$

ere $\varepsilon = 1/bT > 0$. Since we are interested in large bT we consider ε to be all. If $\varepsilon = 0$ equation (5.3) is no longer a differential equation. A problem ch as this, which involves a small parameter ε and which drastically alters s basic form when $\varepsilon = 0$, is called "singularly perturbed". Such problems cur often in applied mathematics and many techniques have been developed to dy them. One of the basic ideas in singular perturbation theory is to view the blem in an appropriate time scale, as determined by the small parameter ε. We e interested in the equation (5.3) for small $t > 0$, and hence it would be sonable to change the time scale in such a way as to magnify small t. Thus, will let $t' = t/\varepsilon^p$ for some constant $p > 0$.

In order to see what an appropriate value of p is, we make the above change variables in (5.3), letting $\overline{N}(t') := N(t'\varepsilon^p)$:

$$\overline{N}'/\overline{N} = \varepsilon^{p-1} f\left(\int_0^{\infty} \overline{N}(t'-s)\overline{k}(s)ds \right)$$

re $\overline{k}(s) = \varepsilon^p k(\varepsilon^p s)$. Note that $\overline{k} \in L^1_+$, $|\overline{k}|_1 = 1$. In order to center the prob- on the equilibrium e let $x(t') = \ln(\overline{N}(t')/e)$. Then this equation becomes

$$(5.4) \qquad x'(t') = \varepsilon^{p-1} f\left(\int_0^\infty e \exp\,(x(t'-s))\bar{k}(s)ds\right).$$

Suppose we look for solutions near equilibrium of the form

$$(5.5) \qquad x(t') = \varepsilon^p y(t') + \varepsilon^p z(t',\varepsilon^p)$$

where z is higher order in ε^p, that is to say $|z(t',\varepsilon^p)| = 0(\varepsilon^p)$ uniformly for $0 \le t' \le \tau$ for a given fixed $\tau > 0$. The idea is simply to plug (5.5) into (5.4) and equate the resulting coefficients of like powers of ε.

In order to see explicitly the lowest order terms in ε we use the assumption (5.2) on f to write $f(w + e) = f'(e)w + r(w)$, $r(w) = 0(|w|)$. Then (5.4) reduces to

$$(5.6) \qquad x'(t') = \varepsilon^{p-1}\left[-d\int_0^\infty x(t'-s)\bar{k}(s)ds + R(x)\right], \qquad d: = -ef'(e) > 0$$

$$R(x): = -d\int_0^\infty g(x(t'-s))\bar{k}(s)ds + r\left(\int_0^\infty eh(x(t'-s))\bar{k}(s)ds\right)$$

$$g(x): = -1 - x + \exp\,(x), \qquad h(x): = -1 + \exp\,(x).$$

Note that $g(x)$ and hence $R(x)$ are higher order in x. We can write

$$\bar{k}(s): = \varepsilon^p k(\varepsilon^p s) = \varepsilon^p[k(0) + k'(0)\varepsilon^p s + 0((\varepsilon^p s)^2)]$$

provided we assume that $k(s)$ is twice continuously differentiable at $s = 0$. Then substituting (5.5) into (5.6) we find that the lowest order term on the left hand side is $\varepsilon^p y'(t')$. From the right hand side we obtain the lowest order term

$$(5.7) \quad (a) \qquad -\varepsilon^{3p-1} dk(0)\int_0^\infty y(t'-s)ds \quad \text{if} \quad k(0) \neq 0$$

.7) (b) $$-\varepsilon^{4p-1} dk'(0) \int_0^\infty y(t' - s)ds \quad \text{if} \quad k(0) = 0, \quad k'(0) \neq 0.$$

us, in order for both sides of (5.6) to have lowest order terms of the same der we must choose

$$p = 1/2 \quad \text{if} \quad k(0) \neq 0$$

$$p = 1/3 \quad \text{if} \quad k(0) = 0, \quad k'(0) \neq 0.$$

These two cases are not, of course, exhaustive. If $k(0) = k'(0) = 0$ then smaller value of p must be chosen, depending on the order of the first nonzero rivative of the delay kernel $k(t)$ at $t = 0$. We restrict our attention to the o cases above since they correspond respectively to our "weak" and "strong" neric delay kernels.

(a) <u>If $k(0) \neq 0$ and $p = 1/2$,</u> we let $\theta = \varepsilon^{1/2}$ and substitute $t') = \theta y(t') + \theta z(t',\theta)$, $z = O(\theta)$ into (5.6) which then becomes

.8) $$x'(t') = \theta^{-1}[-d \int_0^\infty x(t' - s)\bar{k}(s)ds + R(x)].$$

is results, to the lowest order in θ, in a linear integrodifferential equation r y:

.9) $$y'(t') + dk(0) \int_0^\infty y(t' - s)ds = 0.$$

e higher order terms yield an equation for $z(t') = z(t';\theta)$:

.10) $$z'(t') + dk(0) \int_0^\infty z(t' - s)ds = R^*(z;\theta)$$

ere $R^*(z;\theta) = O(\theta)$ uniformly for bounded $z(t')$ on $t' \leq \tau$.

Let $y^0(t)$ be a given (initial) function with compact support, i.e. $y^0(t) = 0$ when $t \leq t^*$ for some $t^* < 0$. Define $\bar{y}^0(t') := y^0(t'\theta)$. We wish to solve (5.9) for y and (5.10) for z subject to the initial conditions

$$(5.11) \qquad y(t') = \bar{y}^0(t'), \quad z(t') = 0 \quad \text{for} \quad t' \leq 0.$$

Given $\tau > 0$ it is a straightforward application of the contraction principle to show that (5.11) has a unique solution on $0 \leq t' \leq \tau$ satisfying $z(t') = 0$, $t' \leq 0$ for θ small (for a given solution y of (5.9), which appears in (5.10)). See Cushing (1977b) where more details are given for the similar case (b) (to be considered below).

To solve the linear equation (5.9) for the lowest order term y we rewrite it as

$$y'(t') + dk(0)\int_0^t y(s)ds = d_1$$

$$d_1 := -dk(0)\int_{-\infty}^0 \bar{y}^0(s)ds$$

and then differentiate to find that y must equivalently satisfy the harmonic oscillator equation

$$y'' + \omega^2 y = 0, \qquad \omega = (dk(0))^{1/2}$$

$$y(0) = \bar{y}^0(0), \qquad y'(0) = d_1.$$

Thus,

$$(5.12) \qquad y(t') = \bar{y}^0(0)\cos \omega t' + (d_1/\omega)\sin \omega t'$$

and, to the first order in $\theta = \varepsilon^{1/2}$, x has sustained oscillations of period

$= 2\pi/(-ef'(e)k(0))^{1/2}$.

With regard to the original model (5.1) this means that for large delays T species growth rate responses to changes in its own density ("large" compared the inherent unrestrained growth rate b) the population density tends to hibit undamped oscillations. We say "tends to exhibit" here because first of 1 y in (5.12) is only the lowest order term in the solution and secondly the ove analysis is really only valid on a short time interval (of order $= (bT)^{-1/2}$).

Although we might have expected to see divergent oscillations around equilibum for large delays T, we found (to the first order) sustained periodic oscil-tions. The reason for this is that in this case (a) $k(0) \neq 0$ the delay is eak." In fact, the "weak" generic delay kernel $k(t) = \exp(-t)$ satisfies all the above hypotheses and as we saw in Chapter 3 (cf. Corollary 3) the equilibum is actually asymptotically stable for such a convex, decreasing delay kernel. us, the above analysis should only be interpreted as saying that large delays cause a weakening of the asymptotic stability and a tendency towards oscilla-on in the solutions. It should not be misconstrued so as to imply the existence periodic solutions or even the instability of the equilibrium.

Formally, the results above for this first case (a) can be stated as follows:

THEOREM 5.1 Assume (5.2) holds and that the delay kernel $k(t)$ is twice tinuously differentiable at $t = 0$ with $k(0) \neq 0$. Given any initial function (t) with compact support and given any constant $\tau > 0$ there exists a constant > 0 such that for $bT \geq T_0$ the solution of (5.1) which satisfies

$$N(t) = e \exp(\theta y^0(t)), \quad t \leq 0$$

ere $\theta := (bT)^{-1/2}$ is of the form

$$N(t) = e \exp (\theta y(t/\theta) + \theta z(t/\theta,\theta)) \quad \underline{\text{on}} \quad 0 \leq t \leq \tau\theta$$

where y *is given by* (5.12) *and* $|z(t',\theta)| = O(\theta)$ *uniformly for* $0 \leq t' \leq \tau$.

(b) *If* $k(0) = 0$ *and* $k'(0) \neq 0$ (and hence $p = 1/3$) a completely analogous argument to that for case (a) above can be carried out except with $\theta = \varepsilon^{1/3}$. In this case we would expect stronger oscillations (to the first order) since $k(0) = 0$ means that the delay in the growth rate response is more pronounced than when $k(0) \neq 0$. This turns out to be true in that the first order terms in this case are exponentially divergent. In fact the lowest order term $y(t')$ solves the equation $y''' + dk'(0)y = 0$ together with certain initial conditions related to the initial function $y^0(t)$. The details of this case can be found in a paper by Cushing (1977b).

THEOREM 5.2 *Under the assumptions of Theorem* 5.1 *except that* $k(0) = 0$, $k'(0) \neq 0$ *(and hence* $k'(0) > 0$), *the conclusions of Theorem* 5.1 *remain valid except that* $\theta = (bT)^{-1/3}$ *and* $y(t')$ *is a linear combination of* $\exp(-\lambda t')$, $\exp(\lambda t'/2) \sin \lambda\sqrt{3}\, t'/2$ *and* $\exp(\lambda t'/2) \cos \lambda\sqrt{3}\, t'/2$ *where* $\lambda = (-ef'(e)k'(0))^{1/3} > 0$.

One would be tempted to conclude that the above two theorems say, roughly speaking, that for large values of bT solutions (even starting near equilibrium) exhibit divergent oscillations about the equilibrium. This however would not necessarily be a valid conclusion. The typical example $\exp(-\varepsilon t) \sin t = (1 - \varepsilon t + \ldots) \sin t$ shows that one cannot necessarily neglect higher order terms of some small parameter ε in an expansion valid even for all t and conclude that because the lower order term $(1 - \varepsilon t) \sin t$ exhibits divergent oscillations it follows that $\exp(-\varepsilon t) \sin t$ does also (which it doesn't of course). Nor, as have already pointed out, do these theorems imply that the equilibrium is unable when bT is large.

Thus, no conclusion about the asymptotic behavior as $t \to +\infty$ of solutions of .1) for large bT should be drawn from Theorems 5.1 and 5.2.

In order to see to what extent the first order approximation (with z dropped) Theorem 5.2 is valid for small t the delay logistic

$$N'/N = bT(1 - \int_0^\infty k(s)N(t - s)ds)$$

s numerically solved for the "strong" generic delay kernel $k(t) = t \exp(-t)$. re the equilibrium $e = 1$. FIGURE 5.1 shows two typical cases for "large" values bT; the solid line is the numerically found solution and the dashed line is e computer evaluation of the first order approximation obtained from Theorem 5.2 dropping z. For a given, fixed initial condition, the first order approximaon from Theorem 5.2 was found to be good for larger and larger bT only on a

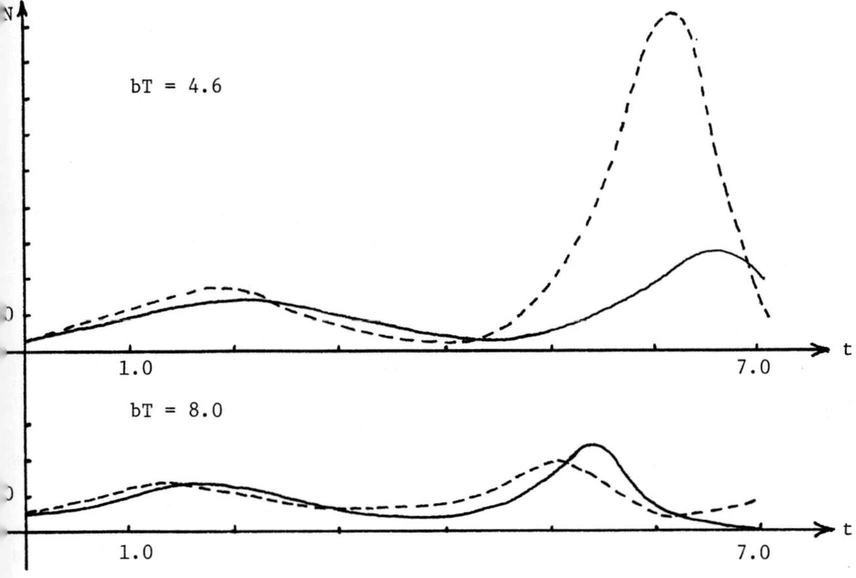

FIGURE 5.1

smaller and smaller t interval as was to be expected. In all cases computed the first order approximation showed oscillations which were more divergent and of smaller frequency than that of the solution itself. Further numerical examples can be found in Cushing (1977b).

5.2 Bifurcation of Periodic Solutions of the Delay Logistic. The simplest example of a single species growth model of the type we have been considering in these notes is the delayed logistic equation

$$(5.13) \qquad N'/N = b(1 - e^{-1} \int_{-\infty}^{t} N(s)k(t-s)ds), \quad b > 0, \quad e > 0$$

for $k \varepsilon L_+^1$, $|k|_1 = 1$, which because all of our analysis has been local (near the equilibrium $N = e$) serves as a prototype for more general delay models. Moreover, the "strong" generic delay kernel $k(t) = T^{-2} t \exp(-t/T)$, $T > 0$ serves as a prototype delay kernel for a model in which the instantaneous growth rate response to density changes at any time t is small and negligible, but in which past population densities exert more and more influence on the growth rate, a maximum growth rate response being to the density at the earlier time $t - T$ with this response depending decreasingly less on even earlier densities. This model with this generic kernel can accordingly be viewed as a more realistic, smoothed or "continuously distributed lag" version of the often studied logistic equation with a constant time lag T.

If we rescale the time variable so that $T > 0$ is the unit of time then (5.13) reduces to

$$(5.14) \qquad N'/N = bT(1 - e^{-1} \int_{-\infty}^{t} N(s)k(t-s)ds), \quad k(t) = t \exp(-t)$$

(where for simplicity we have relabeled time as t). We have seen (Chapter 3, ction 3) that the equilibrium $N \equiv e$ is A.S. for $bT < 2$ and unstable for > 2. Further, the numerical studies by Cushing (1977b) show that as bT nges from small values to large values the solutions of (5.14) show at first sentially monotonic approach to equilibrium, then oscillatory decay to equilibm followed by what are apparently sustained oscillations about equilibrium and en finally violently divergent oscillations about equilibrium (as is consistent th all of our analysis concerned with this equation to this point).

In this section we consider the possibility that stable (nonconstant) period- solutions of (5.14) exist for values of bT at least near the critical value $bT = 2$. Many authors have considered the existence of periodic solutions of layed versions of the logistic. For example, Jones (1962a,b) proves existence eorems for nonconstant periodic solutions of the less realistic, instantaneous ne lag version of (5.14) and Dunkel (1968a,b) and Walther (1975a) prove such eorems for the case when the delay kernel has compact support (their results l be described below). These authors use difficult and lengthy arguments based certain "asymptotic" or "non-ejective" fixed point theorems. Our approach to e prototype equation (5.14) will be by way of the classical Hopf bifurcation eorem, which we will apply after using a trick (which goes back at least to lterra (1909)) to convert (5.14) to a differential system without delay.

Let $x_1 = N - e$. Then (5.14) becomes

$$x_1' = -bT(x_1 + e)e^{-1} \int_{-\infty}^{t} x_1(s)k(t - s)ds. \tag{15)}$$

nsider for the moment the expression

$$x_2(t) := \int_{-\infty}^{t} x_1(s)k(t - s)ds. \tag{16)}$$

cause the "strong" generic kernel with unit delay $k(t) = t \exp(-t)$ satisfies the second order, linear differential equation $k'' + 2k' + k = 0$ with $k(0) = 0$, $k'(0) = 1$, it follows easily that x_2 satisfies the equation

$$x_2'' + 2x_2' + x_2 = x_1. \tag{5.17}$$

Thus, if N is a solution of (5.14) then x_1, x_2 and $x_3 := x_2'$ solve the differential system

$$\begin{aligned} x_1' &= -bT(x_1 + e)e^{-1}x_2 \\ x_2' &= x_3 \\ x_3' &= x_1 - x_2 - 2x_3. \end{aligned} \tag{5.18}$$

Conversely, suppose x_1, x_2 and x_3 are periodic solutions of (5.18). The x_2 is a periodic solution of the linear, second order differential equation (5.17 with periodic forcing term x_1. It is well known from the elementary theory of differential equations that (5.17) has a unique periodic solution x_2 (for period ic x_1) which in fact is the right hand side of (5.16). Thus, x_2 is related to x_1 by (5.16) which implies x_1 solves (5.15) and hence $N = x_1 + e$ is a periodi solution of (5.14).

Consequently, as far as periodic solutions are concerned (5.14) and (5.18) ar equivalent. Moreover, if (5.18) has an orbitally stable A.S. periodic solution, then (since all solutions of (5.14) correspond to solutions of (5.18)) the resulting periodic solution of (5.14) is also orbitally stable. By "orbitally A.S." we mean the following: if $N^*(t)$ denotes the periodic solution, then it is stable (as defined in Chapter 2) and there exists an $\varepsilon > 0$ such that $|N(t) - N^*(t)| \leq \varepsilon$, $t \leq 0$, implies $|N(t) - N^*(t + c)| \to 0$ as $t \to +\infty$ for some phase shift c. Thus, solutions initially near the periodic solution N^* will d to some phase shift of N^*. (It is clear that a periodic solution N^* of 14) could not be A.S. as defined in Chapter 2 since phase shifts of N^* do not d to N^* although they may be initially arbitrarily close to N^*.)

Classical Hopf bifurcation theory (e.g. see Poore (1976)) gives conditions er which (5.18) has a bifurcating branch of orbitally A.S., nonconstant period- solutions. These conditions depend on the eigenvalues of the coefficient ma- x of the linearized system as they are functions of the parameter bT. Since is the critical value of bT we set $\lambda = bT - 2$. Linearizing (5.18) we obtain linear system

$$x_1' = -(\lambda + 2)x_2, \quad x_2' = x_3, \quad x_3' = x_1 - x_2 - 2x_3$$

se coefficient matrix has eigenvalues z given by the roots of the cubic, racteristic equation

$$z^3 + 2z^2 + z + (\lambda + 2) = 0. \qquad (\text{19)}$$

the critical value $\lambda = 0$ this cubic has roots $z = -2, \pm i$. It follows easily m the implicit function theorem that the cubic has root $z_\pm = z_\pm(\lambda)$ for λ ll such that $z_\pm(0) = \pm i$ and $z_\pm(\lambda)$ is differentiable (in fact analytic). Im- cit differentiation of (5.19) evaluated at $z = z_\pm(\lambda)$ yields $(0) = (1 \pm 2i)/10$ and $\text{Re } z_\pm'(0) = 1/10 > 0$.

These facts: that the pair of eigenvalues $z_\pm(\lambda)$ move from the left to right f plane as λ increases through $\lambda = 0$, that they do not do so through the gin $z = 0$ (but through $\pm i$) and that they do not "pause," i.e. $\text{Re } z_\pm'(0) > 0$, ow us to apply the Hopf theorem to (5.18) (see Poore (1976)). This yields us following theorem concerning the prototype delay logistic (5.14).

THEOREM 5.3 *The delay logistic equation* (5.14) *has nonconstant periodic solutions of the following form*

$$N(t) = e + \mu x\left(\frac{t}{1 + \mu\eta(\mu)}, \mu\right) \quad \underline{\text{for}} \quad bT = 2 + \mu\delta(\mu)$$

where δ *and* η *are differentiable real valued functions defined for* $\mu\varepsilon[-\mu_1,\mu_1]$ *for some* $\mu_1 > 0$, *where* $\delta(0) = \eta(0) = 0$ *and where* $x(t,\mu)$ *is, for each* $\mu\varepsilon[-\mu_1,\mu_1]$, *a* (*nonidentically zero*) 2π-*periodic function of* t. *Thus,* $N(t)$ *is* $2\pi(1 + \mu\eta(\mu))$-*periodic in* T. *Moreover, these are the only periodic solutions near equilibrium*.

Consequently the delay logistic (5.14) has nontrivial periodic solutions of approximate period 2π for values of bT near the critical value of 2. The nature of the bifurcation of periodic solutions occurring at $bT = 2$ is described by the two functions δ and η. That is, the period will be near, but smaller or larger than 2π depending on whether $\eta'(0)$ is negative or positive (if it is in fact nonzero). Also, period solutions will exist of bT near, but less than or greater than 2 depending on whether $\delta'(0)$ is negative or positive respectively (if $\delta'(0) \neq 0$). Thus it is of interest to compute these two quantities. Moreover, the orbital stability of the nonconstant periodic solutions turns out to depend on $\delta'(0)$; $\delta'(0) > 0$ implies orbital stability (Poore (1976)).

The quantities $\delta'(0)$ and $\eta'(0)$ can be computed by means of standard perturbation methods or they may be found by means of general formulas available for such Hopf bifurcation problems (see Poore (1976)). Since we do not wish to include the tedious details here, we simply state that it turns out that $\delta'(0) > 0$ and $\eta'(0) > 0$ for our problem (5.18). As a result of this, the *nonconstant periodic solutions of* (5.14) *guaranteed by Theorem* 5.3 *are of period slightly longer than* 2π, *exist for* bT *slightly larger than* 2 *and are orbitally stable*

Thus the prototype delay logistic (5.14) exhibits a typical, standard bifur-
ion phenomenon: as bT is increased through the critical value 2 the equilib-
m e passes from (asymptotic) stability to instability, accompanied by the
earance of stable, nonconstant periodic solutions.

Note that $\delta(0) = 0$, $\delta'(0) > 0$ implies that for a given $bT > 2$ there cor-
spond two periodic solutions from Theorem 5.3, one for $\mu > 0$ and another for
: 0. Actually there are of course infinitely many periodic solutions since any
nslation of a solution of (5.14) is still a solution.

Since any positive, periodic solution is necessarily bounded above and bound-
away from zero it follows from the results of Section 3.1 (ii) that the noncon-
nt periodic solutions found in Theorem 5.3 have average equal to the equilib-
m value e, which in turn implies that they oscillate about e. The same is
n true of the solutions initially near these periodic solutions since the latter
orbitally stable. Suppose for $\mu\varepsilon(0,\mu_1]$ we plot the maximum of the noncon-
nt periodic solution (which is larger than e) against the corresponding value
bT while, for $\mu\varepsilon[-\mu_1,0)$ we plot the minimum. Then we get a typical bifur-
ion diagram as shown in FIGURE 5.2.

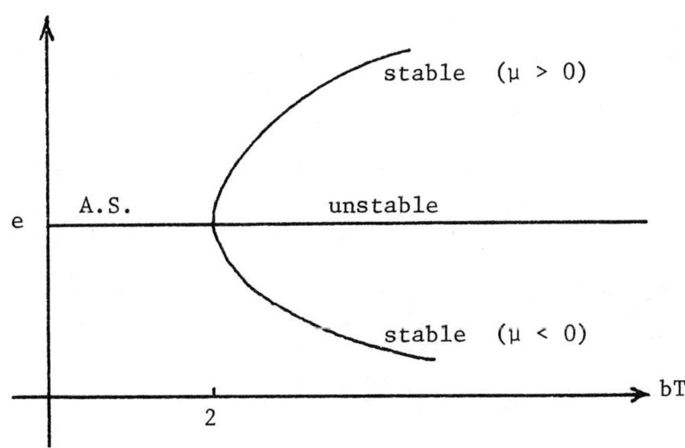

Bifurcation diagram for the delay logistic (5.13).

FIGURE 5.2

5.3 Other Results on Nonconstant Periodic Solutions. The following delay logistic has been studied by Dunkel (1968a,b) and Walther (1975)

$$(5.20) \qquad N'/N = b + m\int_{\gamma}^{t} N(t-s)dh(s), \quad 0 \leq \gamma < \tau < +\infty, \quad b > 0, \quad m > 0$$

who have given existence theorems for nonconstant periodic solutions. Here γ is a "measure" of the delay in the growth rate response in the sense that density changes affect the growth rate only after γ units of time have elapsed. The constant τ is the "maximum life span." The Stieltjes integrator $h(s)$ is <u>decreasing</u> on $[\gamma,\tau]$ with $h(\tau) = 0$ and is possibly discontinuous.

If $h(s)$ is continuously differentiable on $[\gamma,\tau]$ then (5.20) can be written in the form of the delay logistic (5.14) considered in the previous section with

$$k(s) = \begin{cases} -h'(s)/h(\gamma), & \gamma < s < \tau \\ 0, & s < \gamma \text{ and } \tau < s \end{cases}$$

and $e = b/mh(\gamma) > 0$. This delay kernel (which satisfies $k\varepsilon L_{+}^{1}$, $|k|_1 = 1$) is qualitatively similar to the "strong" generic kernel as shown in FIGURE 5.3 for

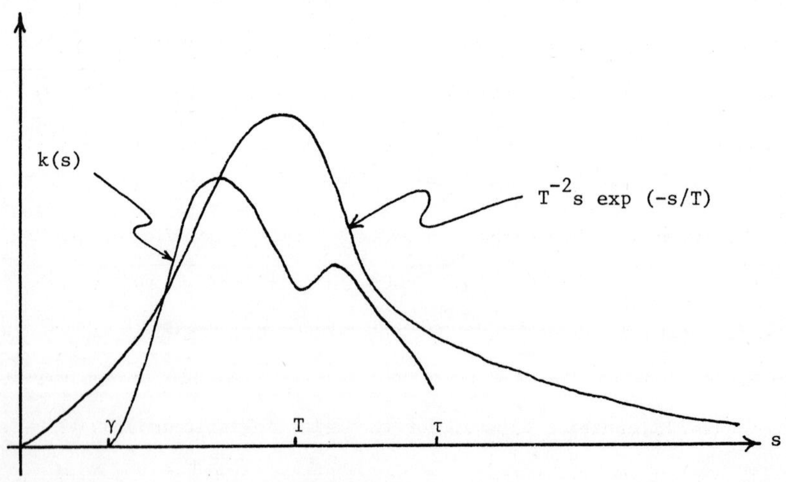

FIGURE 5.3

$\leq T \leq \tau$. Thus, we expect qualitatively similar results for (5.20) and (5.14).

The analysis of (5.20) carried out by Dunkel (1968a) and Walther (1975, 1976) far too complicated to detail here, so we will confine ourselves to the following summary of their results (also see Dunkel (1968b)):

(1) all solutions are bounded: $0 \leq N(t) \leq \max \{N(0); e \exp (b\tau)\}$ for ≥ 0;

(2) if a solution $N(t)$ does not oscillate, then it $N(t) \to e$ as $t \to \infty$ function is said to <u>oscillate</u> if it is neither constant nor monotonic);

(3) if $b\gamma > 1$ then no nontrivial solution is monotonic (i.e. $N' > 0$ or < 0) for all $t > 0$, in other words, all nonconstant solutions oscillate;

(4) if $b\tau \leq 1$ then every solution $N(t) \to e$ as $t \to +\infty$;

(5) if $\gamma \neq 0$ and $b\gamma$ is sufficiently large then there exists a (nontrivial) periodic solution;

(6) if $\gamma = 1$ (this is no loss in generality when $\gamma \neq 0$ as it can always accomplished by a change of time scale) and if $b > \pi/2$ then for some constant > 0 there exists a (nontrivial) periodic solution for each τ: $1 < \tau < 1 + \tau_0$ th range satisfying $0 \leq N(t) \leq e \exp (b\tau)$.

Results (1) - (5) are due to Dunkel (1968a,b) and (6) to Walther (1975).

This list of properties of solutions of (5.20) is qualitatively similar to e properties we obtained above in Section 5.2 and in Chapter 3 for the prototype, lay logistic (5.14). The main difference between our results and (1) - (6) (bedes the greater difficulty and amount of analysis needed to prove (1) - (6)) is at the latter results are global, whereas those obtained for (5.14) are, as peatedly stated, local results which were obtained by local analysis. In order make a few comparisons we might fix the delays T in (5.14) and γ in (5.22), y $T = \gamma = 1$. Our results for (5.14) predict an A.S. equilibrium for $b < 2$, cillations about an unstable equilibrium for $b > 2$ and small amplitude, orbitly stable nonconstant periodic solutions (of a known period) for b near, but eater than 2. Similarly, (1) - (6) predict bounded solutions for all b, (monotonic) asymptotic of the equilibrium for $b \leq 1$, oscillations for $b > 1$, nonconstant periodic solutions (of unspecified period and stability) for sufficiently large b and finally, if the maximum lifespan τ is sufficiently near the delay $\gamma = 1$, nonconstant periodic solutions (of unspecified period and stability) for all b greater than $\pi/2$. The critical value of the inherent birth rate is different in these two sets of results, but qualitatively the models are clearly quite similar.

One of the main shortcomings in the results (1) - (6) is the lack of any statement concerning the stability of the nonconstant periodic solutions. The stability of a periodic solution of an integrodifferential equation is difficult in general to prove. There do not seem to be any general methods which have been successfully and usefully applied to stability questions concerning anything other than constant (equilibrium) solutions. This is not surprising, since such problems are not easily settled even for nondelay, ordinary differential equations. The usual linearization procedure (as discussed in Chapter 2) leads, in the case of a nonconstant periodic solution, to a "nonautonomous" linear integrodifferential equation of the form (2.2) whose stability, in general, cannot be easily deduced from the coefficients in the equation.

There is a well developed theory of Liapunov functions in the stability theory of functional differential equations (Halanay (1966), Hale (1971)) and some such theory for Volterra integrodifferential systems (see G. Seifert (1973)). However, nontrivial applications of this approach are almost nonexistant (Halanay and Yorke (1971)). In any case, Liapunov functions do not seem to have been successfully applied to stability questions concerning the delay models considered here.

One stability result for nonconstant periodic solutions of the very special case of (5.20) with a constant time lag has been given by Kaplan and Yorke (1975). This result, however, does not seem generalizable to either more general delay uations or to systems.

5.4 Periodically Fluctuating Environments. In all of the models considered far we have assumed that all biological and environmental parameters are con- ant in time. For example, the inherent growth rate and the carrying capacity of species were always assumed to be constants. Mathematically, this means that e models have been "autonomous" in that time t has not appeared explicitly in e integrodifferential equations or, more specifically, that time translations of lutions are solutions. Any biological or environmental parameter, however, is turally subject to fluctuations in time and if a model is desired which takes to account such fluctuations then the model must be nonautonomous. Nonautono- us equations are, of course, more difficult to study in general. They may not ssess equilibria and even if they do the study of nonautonomous linearized dels cannot be carried out algebraically as above by means of some characteris- c equation. One must of course ascribe some properties to the time dependence the parameters in the model, for only then can the resulting dynamics be stud- d accordingly. For example, one might assume they are "nearly" constant or riodic, asymptotically periodic, almost periodic, etc.

We will confine our attention here to the case that the biological or envi- nment parameters are periodic of some common period p and consider the ques- on of the existence of periodic solutions of period p and of their stability. ch solutions play the role played by the equilibrium (or carrying capacity) of e autonomous models considered earlier. The periodic oscillation of the param- ers seems reasonable in view of any seasonal phenomena to which they might be bjected, e.g. mating habits, availability of food, weather, harvesting, etc.

As a nonautonomous generalization of the delay logistic considered in Chapter we consider the model

$$N' = N(b(t) - a(t)N - \int_0^\infty N(t-s)k(t,s)ds + r(t,N)) \tag*{.21)}$$

where the t dependence on the right hand side is p-periodic and r is "higher order" in N. The problem is to find conditions under which (5.21) has a positive, p-periodic solution. Our point of view will be that such solutions should bifurcate from the trivial solution $N \equiv 0$. The motivation for this can be found in the simple, autonomous logistic equation $N' = N(b - aN)$ which has a positive, p-periodic (equilibrium) $N = b/a$. This equilibrium as a function of the inherent birth rate b bifurcates from $N \equiv 0$ at $b = 0$. In the more general nonautonomous model (5.21) we will use essentially (but not exactly) the average $A(b) := p^{-1} \int_0^p b(t)dt$ of the time varying inherent growth rate $b(t)$ as the bifurcation parameter.

Let $P(p)$ be the Banach space of continuous, p-periodic functions under the supremum norm $|N|_0 := \max_{0 \leq t \leq p} |N(t)|$. The following basic hypotheses will be assumed about the linear terms in the growth rate response in model (5.21).

$$(5.22)\quad \begin{cases} \text{Assume } a(\cdot),\ b(\cdot) \varepsilon P(p) \text{ and } k(\cdot,s) \varepsilon P(p) \text{ for } s \geq 0 \text{ with} \\ \int_0^\infty |k(t,s)|ds \leq k_0 < +\infty \text{ for some constant } k_0 > 0 \text{ and all } t \geq 0. \\ \text{Also } a(t) \geq 0,\ k(t,s) \geq 0 \text{ (but not both identically zero).} \end{cases}$$

The remainder term r in (5.21) will be assumed to satisfy

$$(5.23)\quad \begin{cases} r(t,\cdot):\ P(p) \to P(p) \text{ in such a way that} \\ |r(t,N_1) - r(t,N_2)| \leq L|N_1 - N_2|_0 \text{ for some constant } L > 0 \text{ and all} \\ \text{small } N_i:\ |N_i|_0 \leq \delta, \text{ for some } \delta > 0. \end{cases}$$

THEOREM 5.4 <u>Under the hypotheses</u> (5.22) <u>and</u> (5.23) <u>the nonautonomous single species model</u> (5.21) <u>has a positive</u>, p-<u>periodic solution for every time varying</u>, p-<u>periodic inherent growth rate</u> $b(t)$ <u>with a small</u>, <u>positive average</u> $A(b) = p^{-1} \int_0^p b(s)ds$.

We will only sketch the proof of Theorem 5.4. The details of a more general sult (using a slightly different approach) are given by Cushing (1978).

Proof. Let $b\varepsilon P(p)$ be given and let $\lambda := A(b) + 1$. Set $t) := b(t) - A(b) - 1\varepsilon P(p)$. Then $b(t) = \lambda + c(t)$ where $A(c) = -1$. We look r a p-periodic solution of (5.21) as an expansion in a small parameter ε

$$N = \varepsilon x + \varepsilon y(\varepsilon) \quad \text{for} \quad \lambda = \lambda_0 + \mu(\varepsilon)$$

$$|y(\varepsilon)|_0 = 0(|\varepsilon|), \quad |\mu(\varepsilon)| = 0(|\varepsilon|).$$

these substitutions are made into (5.21), then to the first order in ε we nd that $x' = (\lambda_0 + c(t))x$. In order for $x\varepsilon P(p)$ we must choose $\lambda_0 = 1$ (so at $A(\lambda_0 + c(t)) = 0)$ in which case

.24) $$x(t) = x_0 \exp \left(\int_0^t b(s)ds - A(b)t\right), \quad x_0 = \text{constant} \neq 0.$$

The higher order terms in ε yield

.25) $$y' = (\lambda_0 + c(t))y + (x + y)\mu - \varepsilon F(t,y;\varepsilon)$$

$$F(t,y;\varepsilon) := a(t)(x + y)^2 + (x + y)\int_0^\infty (x(s) + y(s))k(t,s)ds$$

$$- \varepsilon^{-1}(x + y)r(t,\varepsilon x + \varepsilon y),$$

l equation to be solved for p-periodic y. We briefly indicate how this can be ne. In order that (5.25) be solvable in $P(p)$ it is necessary that the last terms on the right hand side be orthogonal to the linear adjoint, p-periodic solution $1/x$. Thus, given $y^* \varepsilon P_0(p) = \{y \varepsilon P(p): A(xy) = 0\}$ we choose

$$\mu = \varepsilon A(F(t,y^*;\varepsilon)/x)/(1 + A(y^*/x)) \tag{5.26}$$

in which case (5.25) has a unique solution $y \varepsilon P_0(p)$ when y^* is substituted for y in the last two terms on the right hand side. This sets up a map εG: $P_0(p) \to P_0(p)$ defined by $y = \varepsilon G y^*$, a fixed point of which solves (5.25). Note that μ is well defined by (5.26) for $|y^*|_0$ small. Without going into details here, we can show that G is a contraction mapping on a neighborhood of the origin $y^* = 0$ for ε small enough. To do this we use (5.23); the operator has a factor of ε because μ does. Clearly $|y|_0 = 0(|\varepsilon|)$.

Note from (5.26) with $y^* = y$, the fixed point, that

$$\mu = \varepsilon A(F(t,0;0)/x) + \ldots = \varepsilon A(ax + \int_0^\infty x(s)k(t,s)ds) + \ldots .$$

Thus, sign μ = sign εx_0 for ε small. From (5.24) we have sign N = sign εx_0 = sign μ. In terms of the original coefficients this means sign N = sign $A(b)$. $\square$

Whereas the simple autonomous logistic equation $N' = N(b - aN)$ has a positive equilibrium b/a for all $b > 0$, Theorem 5.4 guarantees the existence of a positive, periodic solution only for small inherent growth rate averages $A(b)$. By making use of very general results in bifurcation theory (Rabinowitz (1971)), one can show that the branch of solutions $(\lambda,N)\varepsilon R\times P(p)$ found in the proof above exists globally (as a continuum) and "connects to ∞" in the sense that either λ or $|N|_0$ is unbounded (or both). If p-periodic solutions of (5.21) can be shown to have an a priori estimate in terms of λ which has the property that if $|N|_0$ is unbounded then so is λ, then clearly λ must be unbounded. Using this kind of argument it can be shown (Cushing (1978)) that <u>if</u> $a(t) > 0$, $0 \le t \le p$, nd *(in addition to* (5.22) *and* (5.23)) r *is nonpositive*: $r(t,N) \leq 0$, $0 \leq t \leq p$, or all $N \varepsilon P(p)$ *then* (5.21) *has a positive*, p-*periodic solution for any* -*periodic inherent growth rate* $b(t)$ *with a positive average* $A(b) > 0$. The a riori estimate of p-periodic solutions is obtained as follows. Let t' be such hat $N(t') = |N|_0$. Then $N'(t') = 0$ and from (5.21) follows that

$$0 \leq \lambda + c(t') - a(t')|N|_0 \quad \text{or} \quad |N|_0 \leq (|\lambda| + |c|_0)/m$$

here $m = \min_{0 \leq t \leq p} a(t) > 0$.

We conclude this section with a brief discussion of the stability of the ositive periodic solutions found in Theorem 5.4. From what we have learned in hapter 3 concerning the autonomous models and the stability of their equilibria, e cannot in general expect the periodic solutions of the nonautonomous model 5.21) to be A.S. unless the delay is in some sense small.

Suppose then that we replace the kernel k by δk^* and the remainder term by δr^* in (5.21) where $\delta > 0$ is to be thought of as a small constant. Let $= \overline{N}(\delta)$ be a positive, p-periodic solution as guaranteed by Theorem 5.4, which an be rather easily shown (by carefully following the details of the proof above) be continuous in δ in the sense that $\overline{N}(\delta) \to \overline{N}(\delta_0)$ in the norm $|\cdot|_0$ as $\to \delta_0$. Thus, if we restrict $\delta \leq \delta_0$ for $\delta_0 > 0$ then $|\overline{N}(\delta)|_0 \leq N_0$ for some onstant $N_0 > 0$ independent of δ. Moreover, if $\delta = 0$ then $\overline{N}(0)$ is a posi-ve periodic solution of the logistic $N' = N(b(t) - a(t)N)$ and as a result is ounded away from zero. This means that by choosing δ_0 smaller if necessary we y assume that all $\overline{N}(\delta)$ are, uniformly in δ, bounded away from zero in time, y by a constant n_0. To summarize, with k and r replaced by δk^* and δr^* (5.21) the p-periodic solutions of Theorem 5.4 satisfy $0 < n_0 \leq N(t) \leq N_0$ for l t and all $\delta \varepsilon [0,\delta_0]$ for some constants $n_0 > 0$, $N_0 > 0$ depending on a ven $\delta_0 > 0$.

Suppose now that

$$(5.27)\quad \begin{cases} r^* \text{ satisfies (5.23) with } P(p) \text{ replaced by the larger space of} \\ \text{functions continuous and bounded for all } t \text{ under the supremum norm} \\ |N|_0 = \sup_t |N(t)|. \end{cases}$$

With these preliminaries out of the way we set $z = (N - \bar{N})/\bar{N}$ where $\bar{N}$ is a positive, p-periodic solution of (5.21). Then from (5.21) it follows that z satisfies an equation of the form

$$z' = (-a(t)\bar{N}(t))z + \delta f(t,\bar{N};z) \tag{5.28}$$

where f contains linear and higher order terms in z and in fact is (uniformly in δ) linearly bounded in z in the sense that $|f(t,\bar{N};z)|_0 \leq \gamma_1 |z|_0$ for all $|z|_0 \leq \gamma_2$, uniformly in $\delta \varepsilon [0,\delta_0]$, for some constants $\gamma_i > 0$.

If $a(t) > 0$ for $0 \leq t \leq p$ (and hence is bounded away from zero) then, the linear equation $z' = (-a\bar{N})z$ is uniformly A.S. uniformly in $\delta\varepsilon[0,\delta_0]$ or more specifically all solutions satisfy $|z(t)| \leq \gamma_3 \exp(-\gamma_4 t)$ for some constants $\gamma_i > 0$, all $t \geq 0$ and all $\delta\varepsilon[0,\delta_0]$. This fact together with the factor of δ on the perturbation term in (5.28) allows us to conclude that, for δ_0 small, $z = 0$ is uniformly A.S. as a solution of (5.28). (This can be proved by simple modifications of basic perturbation theorems for ordinary differential equations. For example, see Theorem 8 and its proof in Coppel's book (1965).)

Thus: _if_ $k = \delta k^*$ _and_ $r = \delta r^*$ _in_ (5.21) _where_ k^* _satisfies_ (5.22) _and_ r^* _satisfies_ (5.23) _and_ (5.27), _then the_ p-_periodic solutions of_ (5.21) _guaranteed by Theorem_ 5.4 _are_ (_locally and uniformly_) _A.S. for_ δ _sufficiently small_.

Thus, for delays of small magnitude the periodic, delay logistic (5.21) has a positive, periodic and A.S. solution which accordingly plays the role of a time arying (periodic) carrying capacity for a species in a periodically fluctuating nvironment.

As the delay becomes more significant we would expect the stability of this eriodic solution to be lost, as is the case for the autonomous model. This probem, to the author's knowledge, has not been studied.

CHAPTER 6. OSCILLATIONS AND MULTI-SPECIES INTERACTIONS WITH DELAYS

The main purpose in this chapter is to explore the possibility that sustained oscillations about an equilibrium can be caused by delays in growth rate responses of one or more members in a multi-species community. Taking a hint from the single species case studied in the previous Chapter 5, we will study this question as a bifurcation problem using the inherent, unrestrained growth rate b_i of each species as system parameters or equivalently, after choosing a time scale in which the "delay" T in the system is of unit length, using the dimensionless parameters b_iT. A typical case of the type we will investigate would be that of a community of two or more species whose stable equilibrium densities become unstable as the parameter b_iT of one or more species passes through a critical value at which point the species densities then exhibit periodic oscillations about the now unstable equilibrium. We will give a method to find these critical values as well as the period of the resulting oscillation.

In Section 6.1 a general bifurcation theorem is given (without proof) which is applicable to the ecological models being considered here. In following sections applications to specific models are considered. One feature of the approach taken in these sections is that, unlike the Hopf bifurcation theory used in Section 5.2 for the single species case, more than one parameter b_iT is used in the bifurcation analysis. One result of this is the possibility (especially for communities with a large number n of member species) of not just a single nonconstant, periodic oscillation (stable or unstable), as is the case for systems to which the usual Hopf techniques apply, but of an infinite dimensional manifold of periodic oscillations of a continuum of periods. One shortcoming of our approach below, however, is that we do not obtain any stability results for the periodic solutions.

6.1 A General Bifurcation Theorem. We will present and discuss in this secion a bifurcation theorem of Cushing (1977c) without giving formal, rigorous roofs. In later sections this theorem will be applied to specific ecological odels.

Consider the (Stieltjes) integrodifferential system

$$x'(t) = \lambda_\wedge\left(\int_0^\infty dH(s)x(t-s) + g(x)(t)\right) \tag{6.1}$$

here x and g are n-vectors, H is an $n \times n$ matrix of integrators and λ s a constant <u>n-vector</u>. The operation denoted by "$\wedge$" is defined in the following way: $v_\wedge w = \text{col}\,(v_j w_j)$ where $v = \text{col}\,(v_j)$ and $w = \text{col}\,(w_j)$. Thus, the ight hand side of the ith equation in system (6.1) has a factor λ_i. We wish o consider the case when g is "higher order" in x near $x = 0$ and give conitions under which (6.1) has nontrivial (i.e. $x \not\equiv 0$) periodic solutions for ertain vectors λ. The following basic smoothness assumptions on H and g will e in force throughout:

(H1) $\left\{\begin{array}{l} \text{the entries in } H(s) = (h_{ij}(s)) \text{ satisfy } \int_0^\infty |dh_{ij}(s)| < +\infty \text{ (i.e., are} \\ \text{of finite total variation on } R^+) \text{ and } g(\cdot): B(\rho) \to P(p) \text{ is continuous} \\ \text{for some } \rho > 0 \text{ and } p > 0 \text{ with } |g(x)|_0 = o(|x|_0). \end{array}\right.$

ere $P(p)$ is the Banach space of continuous, p-periodic n-vector valued funcions under the sup norm $|x|_0 = \max_{0 \le t \le p} |x(t)|$ and $B(\rho)$ is the closed all of radius ρ: $B(\rho) := \{x \in P(p): |x|_0 \le \rho\}$.

Before stating the bifurcation theorem we must first briefly discuss the linarization

$$y'(t) = \mu_\wedge \int_0^\infty dH(s)y(t-s) \tag{6.2}$$

and its adjoint

$$(6.3) \qquad z'(t) = -\mu_{\wedge} \int_0^\infty dH^T(s)z(t+s)$$

where H^T is the transpose of H. It can be proved (Cushing (1977c)) that, given a period p, both (6.2) and (6.3) have the same number $r \geq 0$ of independent, p-periodic solutions and that r is finite. This is, of course, for a given n-vector μ. Let $y^{(k)}$ and $z^{(k)}$, $0 \leq k \leq r$, denote r independent, p-periodic solutions of (6.2) and (6.3) respectively. If $v \cdot w = \Sigma v_j w_j$ denotes the usual Euclidean inner product of two n-vectors, let $(y,z) = p^{-1} \int_0^p y(t) \cdot z(t) dt$ and $P_0(p) = \{x \varepsilon P(p): (x, y^{(k)}) = 0, 1 \leq k \leq r\}$. We assume that

(H2) there exists an n-vector μ such that the linearization (6.2) has at least one nontrivial, p-periodic solution (i.e. $r \geq 1$).

Finally, we need to assume something about the $r \times n$ matrix defined by

$$C: = (p^{-1} \int_0^p z_j^{(k)} \sum_{q=1}^n \int_0^\infty y_q(t-s) dh_{jq}(s) dt)$$

where $H(s) = (h_{jq}(s))$. Namely, we assume that

(H3) $n \geq r$ and for a given nontrivial, p-periodic solution $y(t)$ of (6.2) the rank of C is r.

The following theorem has been proved by Cushing (1977c).

THEOREM 6.1 *Under assumptions* H1, H2 *and* H3 *the system* (6.1) *has a nontrivial p-periodic solution of the form*

$$x(t) = \varepsilon y(t) + \varepsilon w(t,\varepsilon) \quad \underline{\text{for}} \quad \lambda = \mu + \gamma(\varepsilon)$$

<u>for all small</u> ε: $0 < |\varepsilon| < \varepsilon_0$ <u>where</u> $w(\cdot,\varepsilon)\varepsilon P_0(p)$ <u>and</u> $|w(\cdot,\varepsilon)|_0 = O(|\varepsilon|)$, $|\gamma(\varepsilon)| = O(|\varepsilon|)$.

This theorem gives conditions under which (6.1) has nontrivial p-periodic solution of small amplitude for λ close to the critical vector μ. The crucial hypotheses are H2 and H3 and they only involve analysis of the linearization (6.2) and its adjoint (6.3). For this reason we now turn to the problem of finding p-periodic solutions of (6.2) and (6.3).

If, in order to find p-periodic solutions of (6.2), we substitute the Fourier series

(6.4) $$y(t) = \sum_{-\infty}^{\infty} c_m \exp(im\omega t), \quad \omega = 2\pi/p$$

into (6.2) and equate resulting coefficients of like exponentials on both sides of this equation, then we will find that the complex Fourier coefficients c_m, for $m \geq 0$, must satisfy the algebraic problems

(6.5) $$(\mu \circ H_m(\omega) - im\omega I)c_m = 0, \quad m \geq 0$$

$$H_m(\omega) := \int_0^{\infty} dH(s) \exp(-im\omega s).$$

(The reason these algebraic problems need only be considered for $m \geq 0$ is because it turns out that $c_{-m} = \bar{c}_m$, $m \geq 0$ where the bar "$\bar{\ }$" denotes <u>complex conjugation</u>.) Here the operation denoted by "$\circ$" is defined by

$$v \circ A = (v_i a_{ij})$$

where $v = \text{col}(v_i)$ and $A = (a_{ij})$.

Each solution $c_m \neq 0$ of (6.5) yields a complex valued, nontrivial p-periodic solution $c_m \exp(im\omega t)$ of (6.2) which in turn, if $m \neq 0$, yields two independent real valued p-periodic solutions

$$y(t) = \text{Re}\, c_m \exp(im\omega t) \quad \text{and} \quad \text{Im}\, c_m \exp(im\omega t). \tag{6.6}$$

Similarly a substitution of the Fourier series

$$z(t) = \sum_{-\infty}^{\infty} d_m \exp(im\omega t), \qquad \omega = 2\pi/p$$

into the adjoint system (6.3) yields the algebraic problems

$$((\overline{\mu \circ H_m(\omega)})^T + im\omega I)d_m = 0, \qquad m \geq 0. \tag{6.7}$$

Note that the coefficient matrix for this problem is the conjugate transpose of that for (6.5); it is for this reason that (6.3) is adjoint to (6.2). Any nonzero solution $d_m \neq 0$ of (6.7) yields two independent real-valued p-periodic solutions

$$z(t) = \text{Re}\, d_m \exp(im\omega t) \quad \text{and} \quad \text{Im}\, d_m \exp(im\omega t) \tag{6.8}$$

of the adjoint problem (6.3). From linear algebra theory we know that (6.7) is solvable for nonzero d_m for exactly those $m \geq 0$ for which (6.5) is solvable for nonzero c_m.

In our applications we will be interested only in the case of isolated equilibria. If $x \equiv 0$ is an isolated equilibrium of the original system (6.1) then $y \equiv 0$ (and $z \equiv 0$) is an isolated equilibrium of the linearized system (6.2) (and its adjoint (6.3)). This implies that $c_0 = 0$ in the Fourier series (6.3) which n turn implies that the coefficient matrix of (6.5) is nonsingular for $m = 0$, .e. $\det(\mu \circ H_0) \neq 0$. From the definition of the operation "$\circ$" we conclude that $\mu_j \neq 0$ for all $1 \leq j \leq n$. As a result, in this case of an isolated equilibrium, we can simplify the algebra problems (6.5) and (6.7) as well as the hypothesis H3 n Theorem 6.1.

THEOREM 6.2 *Assume* H1 *and that* $x \equiv 0$ *is an isolated equilibrium of* (6.2).

(a) *Then* H2 *holds for some* $p > 0$ *and only if*

$$\det(H_1(\omega) - i\xi \circ I) = 0, \quad H_1(\omega) = \int_0^\infty dH(s) \exp(-i\omega s) \tag{6.9}$$

for some $\xi = \mathrm{col}\,(\xi_j)$, $\xi_i > 0$ *in which case* $\mu_j = \omega/\xi_j$, $\omega = 2\pi/p$.

(b) *If* (6.9) (*i.e.* H2) *holds for some* ξ *then* $r > 0$ *is even*.

(c) *Hypothesis* H3 *holds* if and only *if* $n \geq r$ *and the rank of the* $r \times n$ *matrix*

$$C^*: = \left(p^{-1} \int_0^p z_j^{(k)}(t) y_j'(t) dt\right)$$

is equal to r.

Proof. (a) Clearly the linear system (6.2) has a nontrivial p-periodic solution if and only if $\det(\mu \circ H_m(\omega) - im\omega I) = 0$ for some $m = m_0 > 0$. (We have already seen that this determinant is nonzero for $m = 0$.) Define $p' = p/m_0$. Then since $m_0\omega = 2\pi/p'$ we see that

$$H_{m_0}(\omega) = \int_0^\infty dH(s) \exp(-im_0\omega s) = H_1(2\pi/p')$$

and hence

$$\mu\circ H_{m_0}(\omega) - im_0\omega I = \mu\circ H_1(\omega') - i\omega' I, \qquad \omega' = 2\pi/p'.$$

The homogeneous algebraic system with this matrix as its coefficient matrix is equivalent to the system

(6.10) $$(H_1(\omega') - i\xi\circ I)c_1 = 0, \qquad \xi = \text{col}\,(\omega'/\mu_j)$$

since the jth equation can be divided by $\mu_j \neq 0$. Thus, without any loss in generality, (6.5) is solvable for nonzero c_m for some $m = m_0 > 0$ and $p > 0$ if and only if (6.10) is solvable.

All of this is really just to say that if one considers the <u>minimal period</u> of a solution then (6.5) is solvable for nonzero c_m for $m = 1$ (and that the jth equation can be divided by $\mu_j \neq 0$).

(b) Let $M^+ = \{m > 0\colon \det\,(H_m(\omega) - im\xi\circ I) = 0\}$. Equation (6.5) is solvable for $c_m \neq 0$ if and only if $m\varepsilon M^+$. Let $R(m)$ be the rank of $H_m(\omega) - im\xi\circ I$ for $m\varepsilon M^+$: $0 \leq R(m) < n$. Then (6.5) has $N(m) = n - R(m)$ independent complex solutions $c_m \neq 0$, $m\varepsilon M^+$, each of which yields two independent p-periodic solutions of (6.2) given by (6.6). Thus,

(6.11) $$r = 2\sum_{m\varepsilon M^+} N(m)$$

(c) Since

$$\sum_{q=1}^{n}\int_0^\infty y_q(t - s)dh_{jq}(s) = y_j'(t)/\mu_j$$

the matrix C reduces to

$$C = \left(p^{-1}\int_0^p z_j^{(k)}(t)y_j'(t)dt/\mu_j\right)$$

hose rank is clearly equal to that of C^*. □

Remark (1) In the case of an isolated equilibrium when, as we have seen, > 0 is even, the assumption that $n \geq r$ in H3 rules out the case $n = 1$. hus, the bifurcation theorem is really only applicable to systems $n \geq 2$.

Remark (2) It turns out that $n - r \geq 0$ of the components γ_j of γ in heorem 6.1 are arbitrary (see Cushing (1977c)). They must still be $O(|\varepsilon|)$, owever. More specifically, C (or C^*) has at least one $r \times r$ nonsingular ubmatrix obtained, say, by eliminating certain columns $J = \{j_1, j_2, \ldots, j_{n-r}\}$. hen γ_j is arbitrary for $j \in J$; for example, we could take $\gamma_j(\varepsilon) = \gamma_j^* \varepsilon$ for rbitrary constants γ_j^*, $j \in J$. The remaining components γ_j, $j \notin J$, are functions f these γ_j^* as well as of ε. Thus, it is possible that a given λ may lie on any bifurcation branches and as a result (6.1) can have many periodic solutions of various periods).

Remark (3) The hypothesis H2 is necessary for the bifurcation of p-periodic olutions of (6.1) from $x \equiv 0$ as described in Theorem 6.1. This is the familiar inearization criterion of bifurcation theory; namely, that bifurcation can only ccur at the "eigenvalues" μ of the linearized problem. As those familiar with ifurcation theory know, bifurcation does not (in general) always occur at such ritical vectors μ and further sufficiency conditions of some kind must hold. uch a sufficiency condition is given by H3 in Theorem 6.1.

Remark (4) Note that system (6.5) for the Fourier coefficients is equivalent o

$$(H_m(\omega) - im\xi \circ I)c_m = 0, \quad \xi = \text{col}\,(\omega/\mu_j)$$

provided all $\mu_i \neq 0$.

In the applications to specific ecological models made in this chapter the equilibrium will be isolated and hence we will use Theorem 6.2 to fulfill the hypotheses H2 and H3 of Theorem 6.1 (the smoothness hypothesis H1 being easily checked by observation). The plan of attack in all of these applications then will be as follows: to find $p > 0$ and $\xi = \text{col}(\xi_j)$, $\xi_j > 0$ such that (6.9) is satisfied, to compute the number r of independent p-periodic solutions of the linearization by (6.11) in order to check that $n \geq r$ and, finally, to compute the rank of C^* and see that it equals r.

6.2 Periodic Oscillations Due to Delays in Predator-Prey Interactions. There is certainly no scarcity of predator-prey models available in the literature which in one way or another give rise to sustained, periodic oscillations of both predator and prey densities. For example, the famous Lotka-Volterra model is well known to have only periodic (positive) solutions. This fundamental model is often criticized because of the marginal nature of its stability (as well as for other reasons) and many other models are put forth which, amongst other things, frequently have the, what is felt to be, more reasonable property of a stable limit cycle. For such models the predator-prey densities would tend asymptotically to a periodic oscillation, independently of their initial states. Moreover, these models unlike the Lotka-Volterra model are structurally stable: the qualitative nature of their dynamics doesn't change when the system is slightly altered (for example, by the addition of a finite prey carrying capacity or the addition of small delays or other higher order interaction terms in the response functions).

In this section we will investigate several predator-prey models with linear response functions f_i to see what extent delays can lead to periodic oscillations in species densities and, in particular, to limit cycles. We will see that if the nherent growth rates, for fixed delay kernel, are near critical values (or equiv-lently if the dimensionless parameters b_iT are near critical values) then the odels will possess limit cycles.

While we consider only models with linear response dependence on densities we oint out that the basic bifurcation Theorem 6.1 essentially is a linearization heorem (the only restraint on the nonlinearity is that it be higher order near uilibrium) in that the crucial hypotheses H2 and H3 deal only with the linearized stem. Thus qualitatively, we really don't lose much by restricting attention to inear response functions in our examples below because, for any other model (say, ne with Holling's response functions as in Chapter 4) with the same lineariza-ions, we will get exactly the same results, allowing only for differences in rameter interpretation.

The first application (a) will be worked out in detail in order to illustrate he use of Theorem 6.1. The details of the remaining applications (b) and (c) ll be sketched or omitted.

(a) Volterra's delay model with prey resource limitation. Consider the edator N_2, prey N_1 model

$$\begin{aligned} N_1'/N_1 &= b_1(1 - N_1/c - c_{12}N_2) \\ N_2'/N_2 &= b_2\left(-1 + c_{21}\int_{-\infty}^{t} N_1(s)\,d_s h(t-s)\right) \end{aligned} \tag{6.12}$$

r $c_{ij} > 0$, $c > 0$ and $b_i > 0$. The positive equilibrium of (6.12) is

$$e_1 = 1/c_{21}, \quad e_2 = \frac{c - 1/c_{21}}{cc_{12}} \tag{6.13}$$

ovided $c > 1/c_{21}$, which we assume is true.

If the model (6.12) is centered on the equilibrium (6.13) by setting

$x_i = N_i - e_i$ we find

$$x_1' = -b_1(x_1 + e_1)(x_1/c + c_{12}x_2)$$

(6.14)

$$x_2' = b_2(x_2 + e_2)c_{21}\int_{-\infty}^{t} x_1(s)d_s h(t - s)$$

which has the form of the general system (6.1) with $\lambda = \mathrm{col}\,(b_1, b_2)$,

$$H(s) = \begin{pmatrix} -e_1 u_0(s)/c & -e_1 c_{12} u_0(s) \\ e_2 c_{21} h(s) & 0 \end{pmatrix}$$

$$g(x)(t) = \mathrm{col}\,(-x_1(x_1/c + c_{12}x_2), c_{21}x_2 \int_{-\infty}^{t} x_1(s)d_s h(t - s))$$

where $u_0(s)$ is the unit step function at $s = 0$. It is clear that hypothesis H1 holds for any period $p > 0$ and $\rho = +\infty$. In order to investigate H2 and H3 we turn to Theorem 6.2 and observe that $x \equiv 0$ is an isolated equilibrium.

First, with regard to H2 we consider the algebraic system (6.10) for the first Fourier coefficient c_1 of y in which

$$H_1(\omega) = \begin{pmatrix} -e_1/c & -e_1 c_{12} \\ e_2 c_{21}[C(1) - iS(1)] & 0 \end{pmatrix}$$

where

$$\int_0^\infty \exp\,(-im\omega s)dh(s) = C(m) - iS(m)$$

$$C(m) := \int_0^\infty \cos m\omega s dh(s), \quad S(m) := \int_0^\infty \sin m\omega s dh(s).$$

Hence, the system (6.9) for the Fourier coefficient $c_1 = \mathrm{col}\,(c_1^1, c_1^2)$ reduces to e system

$$(-e_1/c - i\xi_1)c_1^1 + (-e_1c_{12})c_1^2 = 0$$

$$e_2c_{21}(C(1) - iS(1))c_1^1 + (-i\xi_2)c_1^2 = 0$$

ose coefficient matrix is singular if and only if

$$[e_2c_{12}C(1) - \xi_1\xi_2] + i[\xi_2/cc_{21} - e_2c_{12}S(1)] = 0$$

ich is to be solved for $\xi_i > 0$ $(\xi_i = \omega/\mu_i)$ for some $\omega > 0$ $(\omega = 2\pi/p)$. This clearly possible if and only if ω is such that $C(1) > 0$ and $S(1) > 0$ in ich case

$$\xi_1 = C(1)/cc_{21}S(1), \quad \xi_2 = e_2cc_{12}c_{21}S(1). \tag{.15}$$

der these conditions H2 holds in Theorem 6.1.

Finally we need hypothesis H3. First we note that the algebraic systems (6.5) r the Fourier coefficients of y are equivalent to (after dividing each row by) the 2×2 system with coefficient matrix $H_m(\omega) - im\omega\xi\circ I$, $\xi = \text{col}\,(\omega/\mu_j)$, ee Remark (4), Section 6.1) whose determinant is

$$[e_2c_{12}C(m) - \xi_1\xi_2m^2] + i[m\xi_2/cc_{21} - e_2c_{12}S(m)],$$

ich by the choice of ξ_i in (6.15) reduces to

$$e_2c_{12}[C(m) - m^2C(1)] + e_2c_{12}i[mS(m) - S(1)].$$

us, if

(6.16) for each $m \geq 2$ either $C(m) \neq m^2 C(1)$ or $S(m) \neq mS(1)$

then the only complex nontrivial, p-periodic solution of the linearized model (for $\mu_i = \omega/\xi_i = 2\pi/p\xi_i$) is determined by (6.9); i.e., $y(t) = c_1 \exp(i\omega t)$ where $c_1 \neq 0$ solves (6.9).

These remarks imply first of all that $r = 2$ so that the condition $2 = n \geq r = 2$ is fulfilled under condition (6.16). Secondly, by solving (6.9) for c_1 and using (6.6) we find two independent, real valued p-periodic solutions of the linearized system: $c_1 = \text{col}\,(-e_1c_{12}, e_1/c + i\xi_1)$ and as a result two independent linear periodic solutions are

$$
\begin{aligned}
y^1(t) &= \text{col}\,(-e_1c_{12} \cos \omega t,\ e_1c^{-1} \cos \omega t - \xi_1 \sin \omega t) \\
y^2(t) &= \text{col}\,(-e_1c_{12} \sin \omega t,\ \xi_1 \cos \omega t + e_1c^{-1} \sin \omega t).
\end{aligned}
\tag{6.17}
$$

We take, for the sake of generality, the linear solution $y(t)$ in H3 to be a linear combination of these two solutions: $y(t) = \kappa_1 y^1(t) + \kappa_2 y^2(t)$, $\kappa_1^2 + \kappa_2^2 \neq 0$.

In order to compute C^* we must find the $r = 2$ independent adjoint solutions. This is done by solving (6.7) for $m = 1$, which because its coefficient matrix is singular (it is the conjugate transpose of $\mu \circ H_1(\omega) - i\omega I$) reduces to a single equation for the Fourier coefficient $d_1 = \text{col}\,(d_1^1, d_1^2)$.

$$-\mu_2 e_1 c_{12} d_1^1 + i\omega d_1^2 = 0.$$

Solving this equation for $d_1 = \text{col}\,(i\omega, \mu_2 e_1 c_{12})$ we find two independent adjoint solutions from (6.8) to be

$$z^1(t) = \text{col}\,(-\omega \sin \omega t,\ \mu_2 e_1 c_{12} \cos \omega t)$$

$$z^2(t) = \text{col}\ (\omega \cos \omega t,\ \mu_2 e_1 c_{12} \sin \omega t).$$

: is now a straightforward calculation to find C^*:

$$C^* = \begin{pmatrix} -\kappa_1 e_1 c_{12}\omega^2/2 & \mu_2 e_1 c_{12}\omega[-\xi_1\kappa_1 + e_1 c^{-1}\kappa_2]/2 \\ -\kappa_2 e_1 c_{12}\omega^2/2 & -\mu_2 e_1 c_{12}\omega[e_1 c^{-1}\kappa_1 + \xi_1\kappa_2]/2 \end{pmatrix}.$$

.nce

$$\det C^* = e_1^3 c_{12}^2 \omega^3 \mu_2 c^{-1}(\kappa_1^2 + \kappa_2^2)/4 \neq 0$$

: find that the rank of C^* is $r = 2$ as required.

We summarize the above results in the following theorem.

THEOREM 6.3 *Assume that* $c > 1/c_{21}$ *and* $\int_0^\infty dh(s) = 1$. *Suppose* $p > 0$ *is a* *riod for which*

$$\text{.18)}\qquad C(1) := \int_0^\infty \cos \omega s dh(s) > 0, \qquad S(1) := \int_0^\infty \sin \omega s dh(s) > 0, \qquad \omega = 2\pi/p$$

ld and for which (6.16) *holds. Then there exist* p-*periodic solutions of Vol-* *rra's delay predator-prey model* (6.12) *of the form*

$$N_i(t) = e_i + \varepsilon y_i(t) + \varepsilon w_i(t,\varepsilon) \quad \text{for} \quad b_i = 2\pi/p\xi_i + \gamma_i(\varepsilon)$$

r small ε *where* $y = \text{col}\ (y_i)$ *is any linear combination of the two* p-*periodic* *lutions* (6.17), ξ_i *is given by* (6.15) *and the higher order (in* ε) *terms* $=$ col (w_i), $\gamma = \text{col}\ (\gamma_i)$ *are as in Theorem* 6.1.

Before looking at some special cases for specific delay integrators $h(s)$ we make some remarks concerning this result.

Remark (1) The two hypotheses (6.18) and (6.16), which deal with the period p and the delay integrator $h(s)$, are inequalities. Thus, for fixed $h(s)$, if these hypotheses hold for some period p_0 they will hold for p close to p_0. The critical values $2\pi/p\xi_i$ of the growth rates b_i depend continuously on p and the parameters of the system (6.12).

Remark (2) System (6.12) is autonomous in the sense that time translated solutions are still solutions. Thus, the periodic solutions of Theorem 6.3 may be translated by any amount to yield other periodic solutions. This is the reason why $y(t)$ has two arbitrary constants κ_1 and κ_2 in it.

Remark (3) If it is desired to discuss the dynamics of the predator-prey model (6.12) with respect to the relative time scales $1/b_i$ and the delay T represented by $h(s)$, then a time scale change can be performed as in earlier chapters which make the delay time T equal to the new unit of time. If this is done, then b_i is replaced by b_iT in everything above.

Remark (4) If no delay is present in (6.12) then $h(s) = u_0(s)$, the unit step function at $s = 0$, in which case $\int_0^\infty \sin \omega s dh(s) = 0$ and (6.18) fails to hold. Since, as remarked in Section 6.1, the conditions (6.18) are necessary we conclude that bifurcation of p-periodic solutions as described by Theorem 6.3 cannot occur unless some delay is present in model (6.12).

Remark (5) Note as the inherent prey carrying capacity $c \to +\infty$ that $\xi_1 \to 0$ while $\xi_2 \to 0$ and hence the same is true (in reverse order) for the

:itical bifurcation values μ_i of the growth rates b_i. Thus, there is no compatibility between the above bifurcation result and what we found in Chapter namely, that <u>for fixed</u> b_i the equilibrium of the model (6.12) becomes un-able as $c \to +\infty$.

Example (1) Suppose we consider the case of a single time lag $h(s) = u_T(s)$, e unit step function at $T > 0$. Then (6.18) reduces to

$$.19) \qquad C(1): = \cos 2\pi T/p > 0, \quad S(1): = \sin 2\pi T/p > 0$$

d (6.16) reduces to

$$.20) \quad \cos 2\pi mT/p \neq m^2 \cos 2\pi T/p \quad \text{or} \quad \sin 2\pi mT/p \neq m \sin 2\pi T/p \quad \text{for} \quad m \geq 2.$$

p and T are such that these inequalities hold then bifurcation of p-periodic lutions will occur.

First we note that (6.20) holds. For, suppose both inequalities are equali-es for some $m \geq 2$. Then squaring both sides and adding we get

$$1 = m^4 \cos^2 2\pi T/p + m^2 \sin^2 2\pi T/p$$

, since $\sin^2 2\pi T/p = 1 - \cos^2 2\pi T/p$, for $m \geq 2$

$$0 > 1 - m^2 = m^2(m^2 - 1) \cos^2 2\pi T/p > 0,$$

contradiction.

Thus, bifurcation will occur if condition (6.19) holds. We conclude then that r the predator-prey model (6.12) with a given constant time lag $T > 0$ p-periodic solutions will bifurcate from equilibrium for every period p greater than $4T$ for inherent growth rates b_i close to the critical values $2\pi/p\xi_i$ for ξ_i given by (6.15), namely,

$$\xi_1 = c^{-1}c_{21}^{-1} \cot 2\pi T/p, \qquad \xi_2 = e_2cc_{21}c_{12} \sin 2\pi T/p.$$

Actually (6.19) holds and bifurcation occurs for any period p satisfying $p > 4T$ or $p\varepsilon(4T/(1 + 4n), T/n)$ for some integer $n \geq 1$.

Example (2) Consider model (6.12) with $dh(s) = k(s)ds$ where $k(s)$ is some linear combination of the two generic kernels: $k(t) = (a/T + bt/T^2) \exp(-t/T)$, $T > 0$, $a, b \geq 0$ and $a + b = 1$. We must again fulfill (6.18) and (6.16) for some p.

It turns out that for this kernel

$$C(m) = \frac{a(1 + m^2\omega^2T^2) + b(1 - m^2\omega^2T^2)}{(1 + m^2\omega^2T^2)^2}.$$

$$S(m) = m\omega T \frac{a(1 + m^2\omega^2T^2) + 2b}{(1 + m^2\omega^2T^2)^2} > 0.$$

First, consider the condition (6.16). The function $F(x) := [a(1 + x) + 2b](1 + x)^{-2}$, $x > 0$ has derivative $F'(x) = -[a(1 + x) + 4b](1 + x)^{-3} < 0$, $x > 0$ and hence is strictly decreasing. From this fact follows easily the fact that $S(m) \neq mS(1)$ for $m \geq 2$ and hence (6.16) holds.

Finally, (6.18) holds for some p if and only if $C(1) > 0$, that is $(b - a)\omega^2T^2 < a + b$ or since $\omega = 2\pi/p$ if and only if

$$p^2 > 4\pi^2T^2(b - a)/(a + b) := p_0^2.$$

us, we have two different cases: $b \leq a$, $b > a$. The first case is the case en $k(t)$ is decreasing for $t > 0$ while in the second case $k(t)$ has a positive maximum (see Chapter 1). Thus, the case $b \leq a$ is of a "weak" delay while e case $b > a$ is of a "strong" delay. In the "weakly" delayed case $b \leq a$ we ve the bifurcation of p-periodic solutions for all periods $p > 0$. In the trongly" delayed case $b > a$ we have bifurcation of p-periodic solutions for l periods $p > p_0$. These bifurcations occur of course at the critical values of given by $2\pi/p\xi_i$ and (6.15). See the following Section 6.3 where numerically tegrated examples bear out these calculations.

(b) Volterra's model with response delays in both interspecies interaction rms. The predator-prey model

$$
\text{.21)} \qquad
\begin{aligned}
N_1'/N_1 &= b_1(1 - N_1/c - c_{12}\int_{-\infty}^{t} N_2(s)d_s h_2(t - s)) \\
N_2'/N_2 &= b_2(-1 + c_{21}\int_{-\infty}^{t} N_1(s)d_s h_1(t - s)) \\
&\int_0^{\infty} dh_i(s) = 1
\end{aligned}
$$

rves as a generalization of model (6.12) in the previous application (take $= u_0$, $h_1 = h$). In this model (6.21), the prey growth rate is also allowed to ve a delay in its response to predator density changes.

The details of this application are quite similar to those of the previous plication (a). It turns out in this case that conditions (6.18) and (6.16) are placed by the conditions

$$
\text{.22)} \qquad
\begin{cases}
\Sigma_1(1) > 0, \quad \Sigma_2(1) > 0 \\
\Sigma_2(m) \neq m\Sigma_2(1) \quad \text{or} \quad \Sigma_1(m) \neq m^2\Sigma_1(1) \quad \text{for all} \quad m \geq 2
\end{cases}
$$

where

$$\Sigma_1(m): = C_1(m)S_2(m) + S_1(m)C_2(m), \quad \Sigma_2(m): = C_1(m)C_2(m) - S_2(m)S_1(m)$$

$$C_i(m): = \int_0^\infty \cos m\omega s dh_i(s), \qquad S_i(m) = \int_0^\infty \sin m\omega s dh_i(s).$$

Also it turns out that $\det C^*$ is a nonzero constant multiple of $\Sigma_1(1)$ and hence is nonzero. The following theorem follows from the general bifurcation Theorem 6.1.

THEOREM 6.4 _Assume that the inherent prey carrying capacity_ c _satisfies_ $c > 1/c_{21}$ _in the predator-prey model_ (6.21) _so that the equilibrium_ (6.13) _is positive. If_ $p > 0$ _is a period for which_ (6.22) _holds, then_ p-_periodic solutions of_ (6.21) _bifurcate from equilibrium for growth rates_ b_1, b_2 _close to the cricital values_

$$\mu_1 = 2\pi c c_{21}\Sigma_1(1)/p\Sigma_2(1), \qquad \mu_2 = 2\pi/pce_2c_{12}c_{21}\Sigma_1(1).$$

This theorem generalizes that given by Cushing (1976a).

Example (3) Suppose both prey and predator growth rate responses have a single, instantaneous time lag: $h_i(s) = u_{T_i}(s)$, $T_i > 0$. Then $C_i(m) = \cos 2\pi m T_i/p$ and $S_i(m) = \sin 2\pi m T_i/p$ and

$$\Sigma_1(1) = \sin 2\pi(T_1 + T_2)/p, \qquad \Sigma_2(1) = \cos 2\pi(T_1 + T_2)/p.$$

The second condition in (6.22) can be shown to hold by exactly the same argument as in the previous application (in which $T_2 = 0$) with T replaced by

$+ T_2$. The remaining condition in (6.22) holds and hence p-periodic bifurcation curs at the critical values given in Theorem 6.4 for given lags T_i for periods tisfying $p > 4(T_1 + T_2)$.

Example (4) Cushing (1976a) considers the predator-prey model (6.21) when th response delays are "weak": $dh_i(s) = \exp(-s)ds$. Results from numerical tegrations carried out by computer are also given by Cushing (1976a). These re- lts illustrate, amongst other things, the bifurcation of (a continuous, stable, tracting manifold of) p-periodic solutions for this specific example (see FIGURE 7 in the following Section 6.3).

Since the first model (6.12) considered above (cf. Theorem 6.3) deals with e case when the predator response is more significantly delayed than that of the ey and since the example studied by Cushing (1976a) deals with the case when th predator and prey have similar delays in the growth rate responses to inter- ecies interactions, let us consider here the remaining opposing case of when the ey response is significantly delayed in comparison to that of the predator. wards this end, let

$$dh_2(s) = T^{-2} s \exp(-s/T), \qquad dh_1(s) = \exp(-s)ds,$$

model (6.21). In this case we find that

$$C_1(m) = (1 + m^2\omega^2)^{-1}, \qquad S_1(m) = m\omega(1 + m^2\omega^2)^{-1}$$

$$C_2(m) = (1 - m^2\omega^2T^2)(1 + m^2\omega^2T^2)^{-2}, \qquad S_2(m) = 2m\omega T(1 + m^2\omega^2T^2)^{-2}$$

d as a result

$$\Sigma_1(m) = (1 - m^2\omega^2T^2 - 2m^2\omega^2T)(1 + m^2\omega^2)^{-1}(1 + m^2\omega^2T^2)^{-2}$$

$$\Sigma_2(m) = m\omega(1 + 2T - m^2\omega^2T^2)(1 + m^2\omega^2)^{-1}(1 + m^2\omega^2T^2)^{-2}.$$

First of all note that as $m = 1, 2, 3, \ldots$ increases the expression $\Sigma_1(m)$ decreases and as a result $\Sigma_1(m) \neq m^2\Sigma_1(1)$ for $m \geq 2$. Thus, Theorem 6.4 applies and bifurcation occurs if the two inequalities $\Sigma_1(1) > 0$ and $\Sigma_2(1) > 0$ hold, i.e. if $\omega = 2\pi/p$ and T are such that

$$1 - \omega^2T^2 - 2\omega^2T > 0, \qquad 1 + 2T - \omega^2T^2 > 0.$$

These two conditions hold if $\omega^2 < 1/(T^2 + 2T)$ or

$$p > 2\pi(T^2 + 2T)^{1/2}.$$

(c) Prey response delays to resource limitation. As opposed to the two previously considered predator-prey models of this section in which there are no response delays to intraspecies interactions we next consider the model

$$\text{(6.23)} \qquad \begin{aligned} N_1'/N_1 &= b_1\Big(1 - c^{-1}\int_{-\infty}^{t} N_1(s)\,d_sh_3(t - s) - c_{12}N_2\Big) \\ N_2'/N_2 &= b_2\Big(-1 + c_{21}\int_{-\infty}^{t} N_1(s)\,d_sh_1(t - s)\Big) \\ \int_0^\infty dh_i(s) &= 1. \end{aligned}$$

This is Volterra's delay model with an added delayed logistic term for the prey species. If $h_1(s) = u_0(s)$ so that delays are only present in the prey response to resource limitation, then we have a model considered by May (1973, 1974) (cf.

tion 4.3 above). May, however, only considered the question of the stability the equilibrium. We wish to apply the bifurcation Theorem 6.1 to find conditions under which the more general model (6.23) has nontrivial periodic solutions.

This model (6.23) has equilibrium given by (6.13) so we again assume $1/c_{21}$ in order that this equilibrium be positive.

The analysis of this model (6.24) is similar in detail to that of model (6.12) (a) above. It turns out that the necessary condition H2 for bifurcation in orem 6.1 (see (6.9) in Theorem 6.2) is fulfilled if and only if

$$S_1(1)C_3(1) > 0, \qquad C_1(1)C_3(1) + S_1(1)S_3(1) > 0 \tag{6.24}$$

$$\xi_1 = e_1(C_1(1)C_3(1) + S_1(1)S_3(1))/cS_1(1)$$
$$\xi_2 = e_2cc_{12}c_{21}S_1(1)/C_3(1). \tag{6.25}$$

The sufficiency condition H3 (see Theorem 6.2 (c)) that $n = 2 \geq r = 2$ is isfied provided for each $m \geq 2$

$$\begin{cases} \text{either} & mC_3(m)S_1(1) \neq C_3(1)S_1(m) \\ \text{or} & C_1(m)C_3(1) + mS_3(m)S_1(1) \neq m^2(C_1(1)C_3(1) + S_1(1)S_3(1)). \end{cases} \tag{6.26}$$

ally, the condition $\det C^* \neq 0$ turns out to hold because $\det C^*$ is a nono constant multiple of $S_1(1)$.

THEOREM 6.5 *Assume* $c > 1/c_{21}$ *in the predator-prey model* (6.23). *If* $p > 0$

is a period for which (6.24) and (6.26) hold, then p-periodic solutions bifurcate from equilibrium for inherent growth rates b_i *near the critical values* $2\pi/p\xi_i$ *for* ξ_i *given by* (6.25).

The condition (6.24) or more specifically $S(1) > 0$ rules out the complete lack of delay in the predator response to prey interactions (since $h_1(s) = u_0(s)$ implies $S(1) = 0$). This prevents Theorem 6.5 from being applied to May's model (cf. Section 4.3 above) in which $h_1(s) \equiv u_0(s)$. However, the intent of May's model was to consider the case when prey response delays in prey resource limitation were the most significant delays in the system (see May (1974)), a case which we can study using Theorem 6.4 by choosing $h_1(s)$ to be a "weak" delay integrator as in the following example.

Example (5) Suppose we choose a "strong" prey response delay

$$dh_3(t) = T^{-2}t \exp(-t/T)dt, \qquad T \geq 1$$

and a "weak" predator response delay

$$dh_1(s) = t \exp(-t)dt.$$

Note that we have taken the delay $T \geq 1$. For these delay kernels

$$C_1(m) = (1 + m^2\omega^2)^{-1}, \qquad S_1(m) = m\omega(1 + m^2\omega^2)^{-1}$$

$$C_3(m) = (1 - m^2\omega^2T^2)(1 + m^2\omega^2T^2)^{-2}, \qquad S_3(m) = 2m\omega T(1 + m^2\omega^2T^2)^{-2}.$$

Consequently (6.24) holds if and only if $\omega^2T^2 < 1$ or in other words for periods p for which

$$p > 2\pi T.$$

Finally, we need to check condition (6.26). The first inequality in (6.26) is ivalent to

$$(1 - m^2\omega^2T^2)(1 + m^2\omega^2)(1 + m^2\omega^2T^2)^{-2} \neq (1 - \omega^2T^2)(1 + \omega^2)(1 + \omega^2T^2)^{-2}$$

ch in fact holds for all $m \geq 2$ since the function $(1 - xT^2)(1 + x)(1 + xT^2)^{-2}$, $T \geq 1$, is a strictly decreasing function of $x > 0$.

We conclude that the predator-prey model (6.24) with a "strong" generic delay the prey response to resource limitation and a "weak" delay in the predator ponse to prey contacts has p-periodic solutions which bifurcate from the positive ilibrium of any periods p satisfying $p > 2\pi T$ for b_i close to the critical ues $2\pi/p\xi_i$ for ξ_i given by (6.25).

Before applying the bifurcation Theorem 6.1 to models of other types of interions we describe in the following section some numerical examples which illuste many of the results we have obtained for predator-prey interactions with ays. Also we briefly consider (in Section 6.4) the case of large delays in dator-prey models.

6.3 Numerically Integrated Examples of Predator-Prey Models with Delays. In s section we describe the results of computer integrations of two predator-prey els with generic delays. We will consider Volterra's delay model with first a (nondelay) prey resource limitation term and with a "strong" generic delay the predator response to prey density changes and secondly (b) with "weak" eric delays in both species responses to interspecies interactions. Besides ustrating the bifurcation results of the preceding Section 6.2, these examples ustrate many of the stability and instability results of Chapter 4.

(a) Volterra's delay predator-prey model with inherent prey resource limitation. We wish to consider model (6.12) with the "strong" generic delay kernel $dh(s) = T^{-2}t \exp(-t/T)$, $T > 0$. If (as in Remark (3) following Theorem 6.3) we choose a time scale for which $T > 0$ is the unit of time, then b_iT replaces b_i and $dh(s) = t \exp(-t)$ in the model (6.3) as well as in Theorem 6.3 and in the calculations of Example (2) following Theorem 6.3. We consider then the model

$$(6.27)\qquad \begin{aligned} N_1' &= b_1TN_1(1 - N_1/c - N_2) \\ N_2' &= b_2TN_2(-1 + \int_0^\infty N_1(t - s)s \exp(-s)ds) \end{aligned}$$

$$c > 0, \quad b_i > 0, \quad T > 0$$

where, to be specific, we have chosen the interaction coefficients $c_{12} = c_{21} = 1$. This system has two equilibria in the right half plane:

$$E_1\colon e_1 = c, \quad e_2 = 0 \quad \text{and} \quad E_2\colon e_1 = 1, \quad e_2 = (c - 1)/c.$$

First of all, we have from Theorem 4.4 of Chapter 4 that $c < 1$ implies that all positive solutions tend to E_1. This is illustrated in FIGURE 6.1 where several phase plane trajectories are shown for $c = 0.8$ and $b_1T = 4.0$, $b_2T = 2.0$.

The trajectories in FIGURE 6.1 and in all other FIGURES of this section were computed from initial functions of the form

$$N_1(t) = N_1^0u_0(t), \quad N_2(t) = N_2^0u_0(t), \quad t \leq 0$$

for constants $N_i^0 > 0$ where $u_0(t)$ is the unit step function at $t = 0$; i.e. $u_0(t) = 0$ for $t < 0$ and $u_0(t) = 1$ for $t \geq 0$. Such initial conditions might considered appropriate for a model of two species of densities N_i^0 which are aced together or otherwise begin their interaction at $t = 0$ (without any past story) as, for example, might be the case in certain laboratory experiments.

FIGURES 6.2 and 6.3 show what happens as $c > 1$ is increased from values ar 1 to "large" values (for fixed $b_1T = 4.0$, $b_2T = 2.0$). We see in FIGURE 2 that for $c = 1.2$ and $c = 1.4$ the equilibrium E_2 is A.S. (Theorem 4.5). FIGURE 6.3 we find that for $c = 1.7$ that E_2 is unstable (Theorem 4.5) and e trajectory shows a definite outward spiraling.

Although only a few trajectories are shown, all computed trajectories have e same qualitative and asymptotic properties for equal values of c, as those own in FIGURES 6.1 - 6.3.

The loss of stability of E_2 as c increases (for b_iT are fixed) suggests e existence of a limit cycle for appropriate values of c. This is illustrated FIGURE 6.4 where a computed limit cycle is shown for $b_1T = 4.0$, $b_2T = 2.0$ and = 1.67. Also shown is a trajectory which approaches the limit cycle and thus il- strates its orbital stability. The period of the limit cycle was observed to be proximately $p = 8.5$. The existence of this limit cycle for these parameter ues is consistent with the results of the previous Section 6.2, namely those in ample (2) (with $a = 0$, $b = 1$, $T = 1$) which assert that p-periodic solutions uld exist for

$$28) \qquad p > 2\pi, \quad b_iT \sim \mu_1 := \frac{8\pi^2 c}{p^2 - 4\pi^2}, \quad b_2T \sim \mu_2 := \frac{(p^2 + 4\pi^2)^2}{2(c-1)p^4} .$$

the $c = 1.67$ and $p = 8.5$ these formulae yield $\mu_1 = 4.02$ and $\mu_2 = 1.78$. s, the values $b_1T = 4.0$, $b_2T = 2.0$ used in FIGURE 6.4 are in fact near μ_1, respectively.

FIGURES 6.1 - 6.4 illustrate the nature of solutions of Volterra's model (6.27) a function of the inherent prey carrying capacity c. In FIGURE 6.5 are shown four limit cycles for a fixed value of $c = 2.0$, but for different values of b_iT. Here we chose selected values of μ_1 from which (by (6.28)) was computed the period p and hence μ_2 as shown in the following table. Also listed are the values of b_1T and b_2T for which limit cycles were found as shown in FIGURE 6.5.

μ_1	p	μ_2	b_1T	b_2T
1.0	14.05	0.72	1.0	0.73
2.0	10.88	0.89	2.0	0.93
4.0	8.89	1.13	4.0	1.17
8.0	7.70	1.39	8.0	1.40

One interesting feature of all trajectories which we find from our numerical integrations of Volterra's delay model is related to the unstable (for $c > 1$) equilibrium E_1 on the N_1 (prey) axis. For larger values of (at least one) b_iT the trajectories become more pointed in the direction of E_1. This can be seen in FIGURES 6.4 and 6.5 and is particularly evident in FIGURE 6.6. Moreover, the trajectories spend an inordinate amount of time in the cusp near E_1. This can be explained as follows. Large values of b_iT mean the delay T in predator response is large and as a result the predator initially dies exponentially (as in the total absence of prey) while at the same time the prey density approaches the carrying capacity c. This situation prevails and the trajectory remains near the (unstable!) equilibrium E_1 until enough time, commensurate with the delay, has past for the predator growth rate to respond at which time the trajectory moves away from E_1 (predator density increases). Whether the result is then convergent divergent oscillations (possibly resulting in prey extinction as is suggested FIGURE 6.6) depends on the relative magnitude of the delay T.

Thus, we have found numerically for predator-prey dynamics what was found lytically for single species dynamics in Section 3.6 of Chapter 3: that delays, le generally a destabilizing influence on a stable equilibrium, can be a "stabil- ng" influence on an unstable equilibrium.

It was found that whenever a limit cycle clearly existed in a numerically in- rated example of Volterra's model (6.27) it was apparently unique and globally racting. It is possible, however, in some models that there be more than one it cycle for given values of the system's parameters. This turns out to be the e in the next example.

(b) <u>Volterra's model with delays in both interspecies interaction responses</u>. predator-prey model (6.21) was numerically integrated for $c_{12} = c_{21} = 1$, $= 25$ and for "weak" generic delays $d_s h_i(s) = \exp(-s)ds$. Limit cycles were nd for $b_1 = 14.0$ and $b_2 = 1.0$ as shown in FIGURE 6.7. Also shown are some jectories which orbitally tend to these limit cycles; these often approach so t that they are visible only as short "tails" on the limit cycles in FIGURE 6.7. limit cycles were visibly observed to have period slightly smaller than 7.5 ightly more than four cycles were traced in the time interval $0 \leq t \leq 30$), say 7.4.

These are commensurate with the formulas of Theorem 6.4 which yield

$$\mu_1 = 20\pi^2/(p^2 - 4\pi^2), \qquad \mu_2 = (p^2 + 4\pi^2)^2/3p^4$$

as a result $\mu_1 \sim 12.92$, $\mu_2 \sim 0.99$ for $p \sim 7.4$. Thus we should find p- iodic solutions of periods close to 7.4 for values of b_1 and b_2 near 12.92 0.99 respectively.

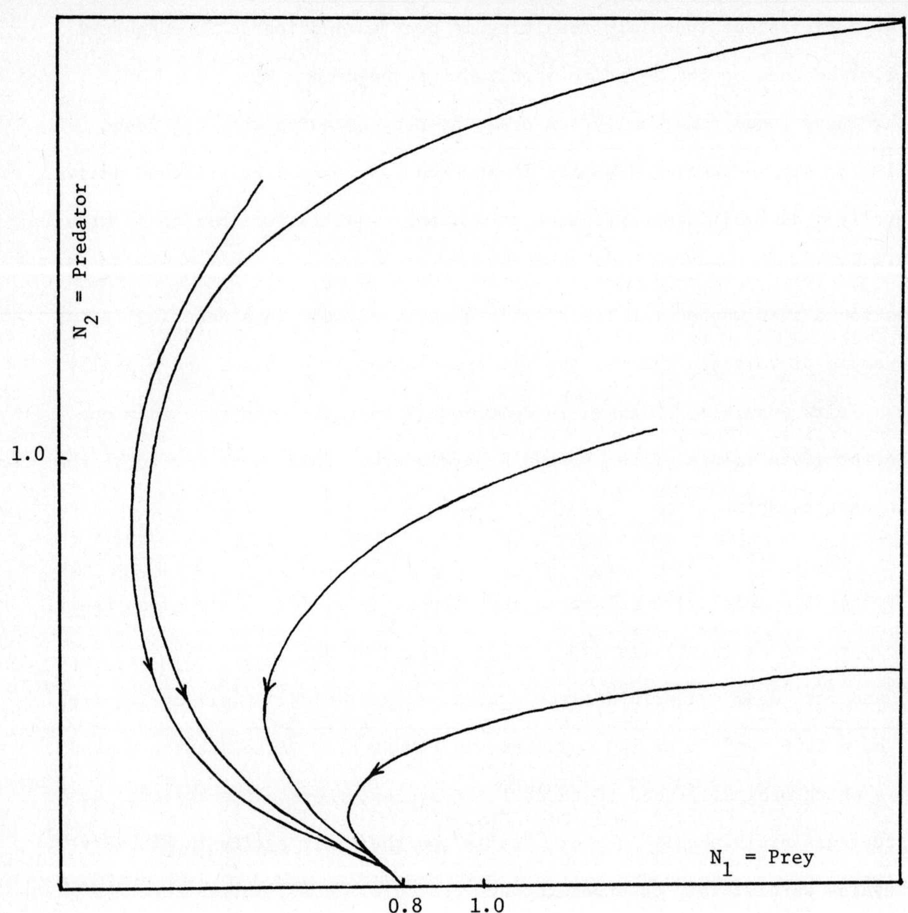

Four solution trajectories of the delay predator-prey model (6.27) are shown for $b_1T = 4.0$, $b_2T = 2.0$ and $c = 0.8$. These illustrate the global attractivity of the equilibrium E_1: $e_1 = 0.8$, $e_2 = 0.0$ when the inherent carrying capacity c is less than 1.0.

FIGURE 6.1

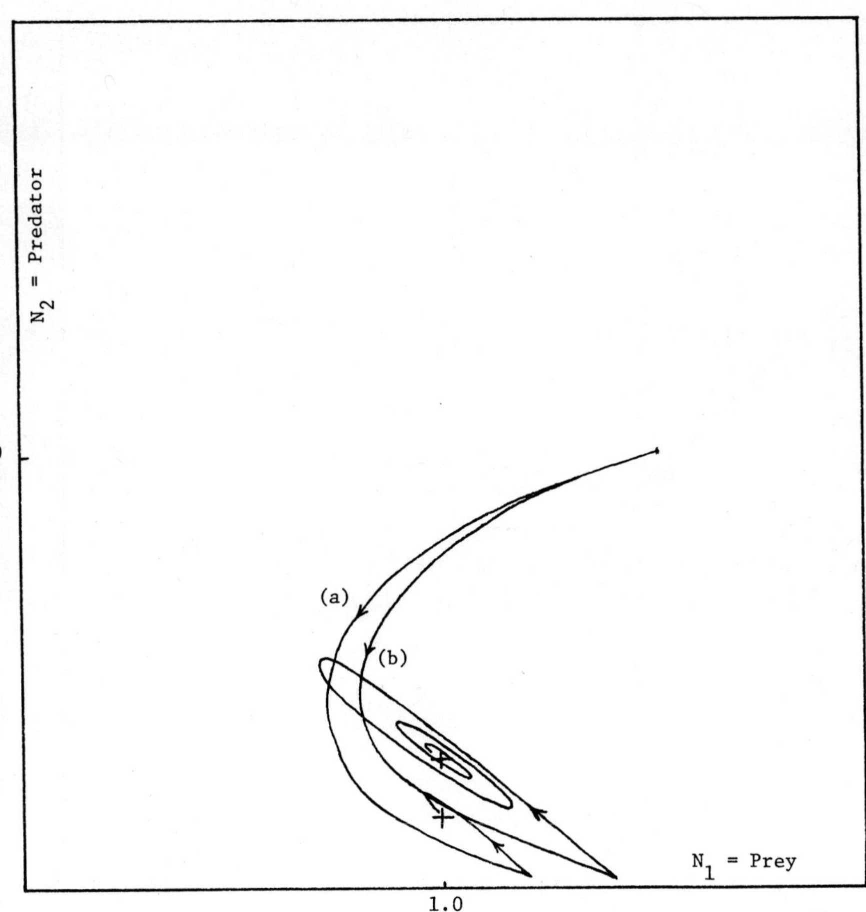

Two trajectories (with the same initial data) of the predator-prey model (6.27) for $b_1T = 4.0$, $b_2T = 2.0$ demonstrate the asymptotic stability of equilibrium E_2: $e_1 = 1.0$, $e_2 = (c - 1)/c$ when $c > 1.0$ is close to 1.0. Trajectory (a) is for $c = 1.2$ and (b) is for $c = 1.4$.

FIGURE 6.2

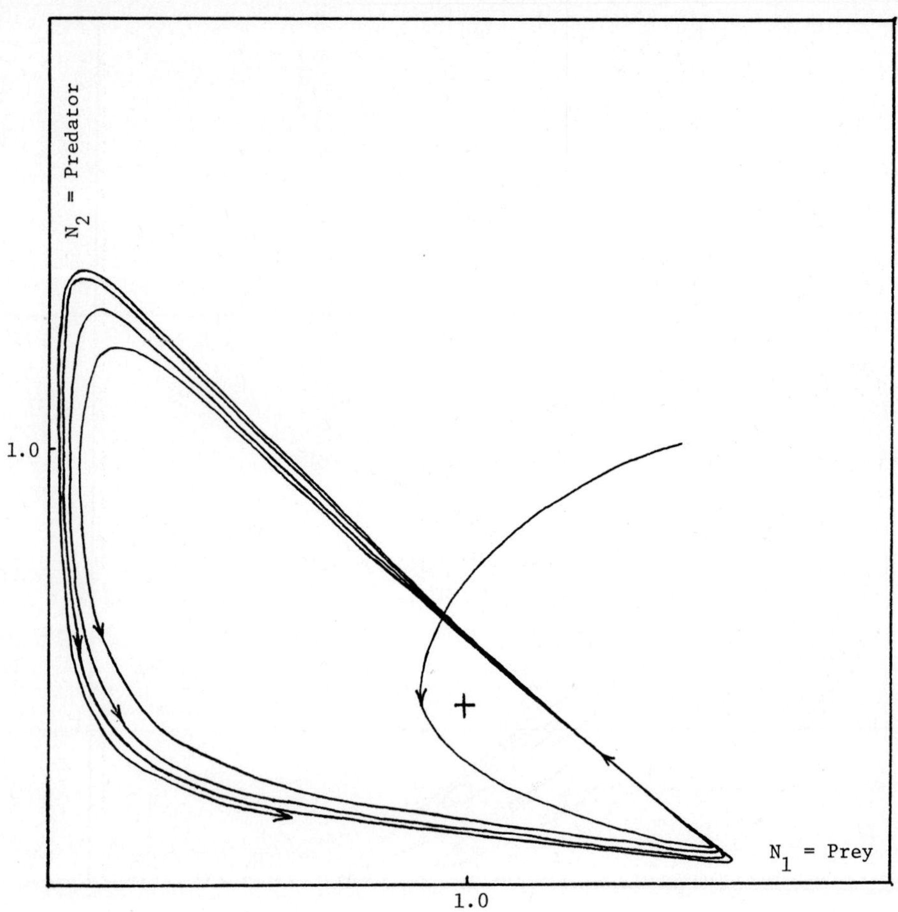

An unstable trajectory (with the same initial data as the stable trajectories in FIGURE 6.2) of the predator-prey model (6.27) with $b_1T = 4.0$, $b_2T = 2.0$, $c = 1.7$ and equilibrium E_2: $e_1 = 1$, $e_2 = 0.41$.

FIGURE 6.3

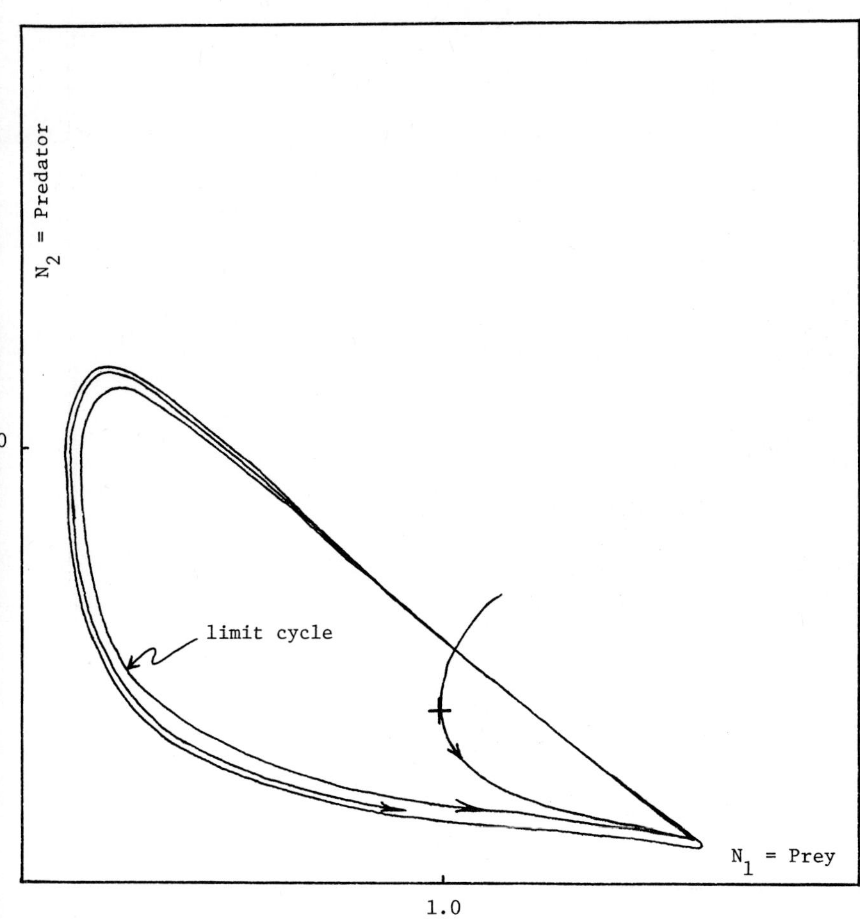

A limit cycle is shown for the predator-prey model (6.27) with $b_1T = 4.0$, $b_2T = 2.0$ and $c = 1.67$. Also shown is a trajectory spiralling inwardly to this limit cycle.

FIGURE 6.4

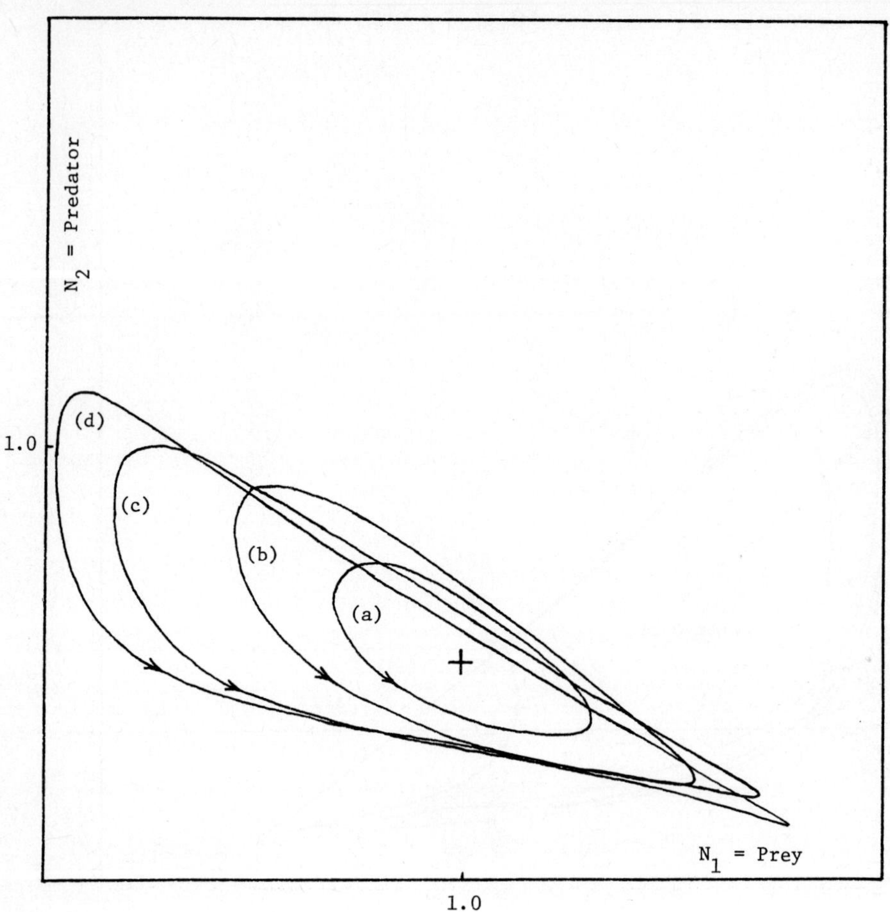

Four limit cycles of the predator-prey model (6.27) with $c = 2.0$ are shown for selected values of the dimensionless parameters b_iT: (a) $b_1T = 1.0$, $b_2T = 0.73$; (b) $b_1T = 2.0$, $b_2T = 0.93$; (c) $b_1T = 4.0$, $b_2T = 1.17$ and (c) $b_1T = 8.0$, $b_2T = 1.4$.

FIGURE 6.5

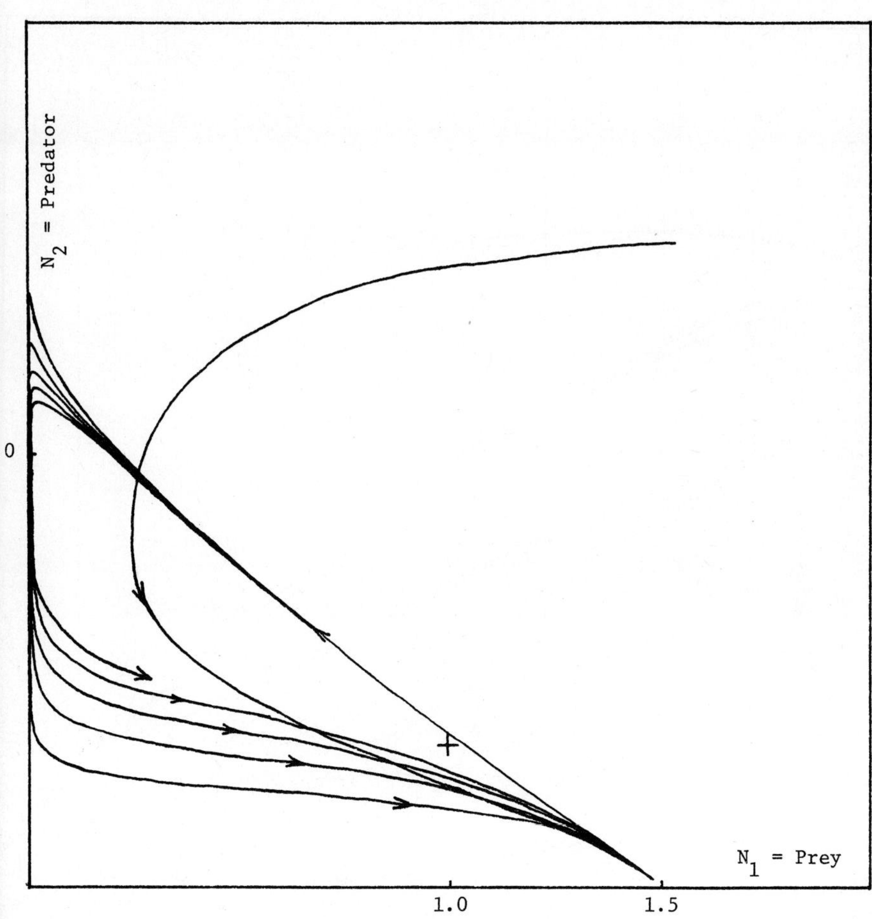

A single trajectory of the predator-prey model (6.27) for $b_1T = 30.0$, $b_2T = 4.0$ and $c = 1.5$ is shown. The trajectory is inwardly spiralling towards the equilibrium E_2: $e_1 = 1$, $e_2 = 1/3$. Besides exhibiting a cusp-like point near the equilibrium E_1: $e_1 = 1.5$, $e_2 = 0$, the trajectory spent the greater portion of the time needed for a cycle near this cusp.

FIGURE 6.6

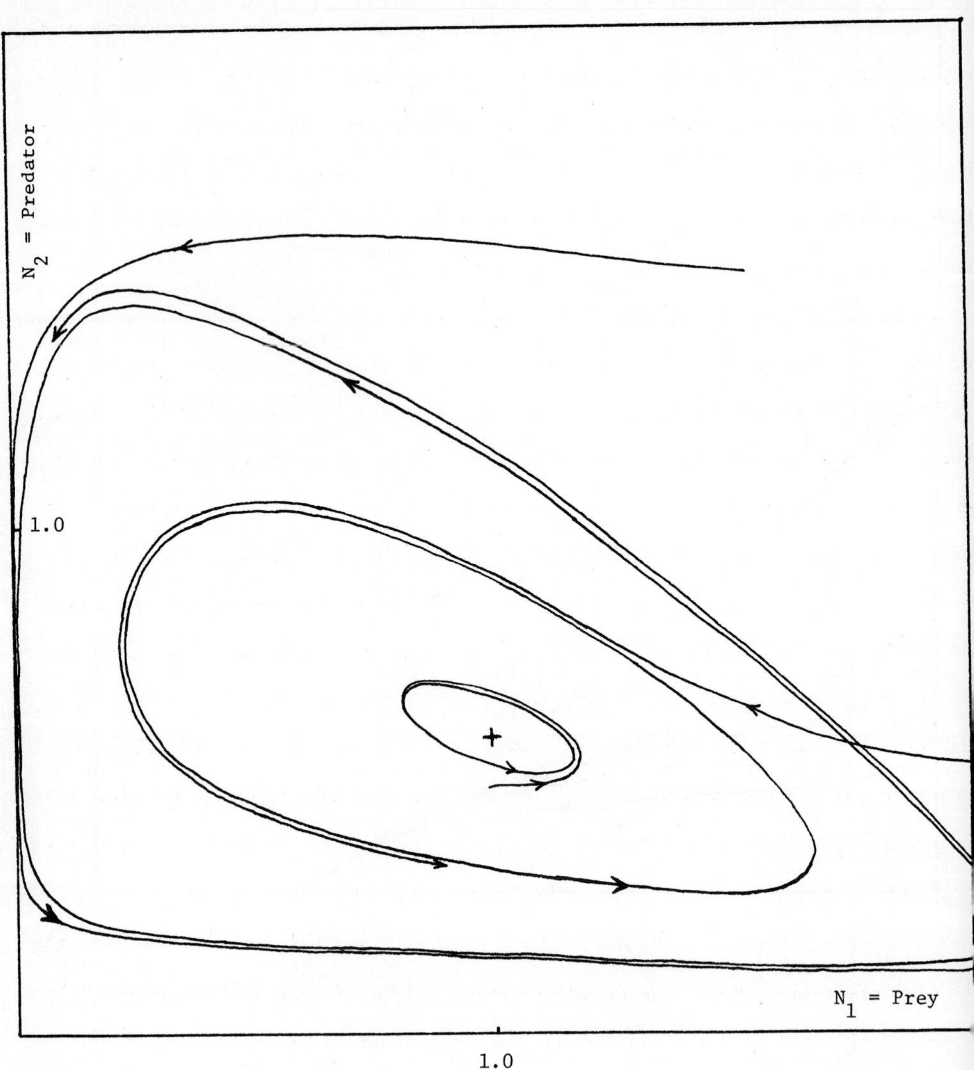

Three limit cycles, each with an approaching trajectory, are shown for the predator-prey model (6.21) with $b_1 = 14.0$, $b_2 = 1.0$, $c_{12} = c_{21} = 1.0$, $c = 2.5$ and "weak" generic delay kernels.

FIGURE 6.7

As in the previous Example (a) above large values of the inherent prey carry-ng capacity c yield an unstable equilibrium E_1 while c near 1 (but $c > 1$) ield E_1 to be A.S. For $c < 1$, E_2 is globally attracting. These facts were bserved in the numerical integrations but since the graphs are similar to those bove in FIGURES 6.1 - 6.6 for the previous example we do not reproduce any here. ee Cushing (1976a) for further computer solutions of this model.

Note in FIGURE 6.7 that several periodic solutions exist. It was found from ur numerical integrations that all initial states in this case gave rise to rajectories which orbitally approached a limit cycle, but not all to the same one s was the case in the previous example. This might be explained as follows: the eneral bifurcation result Theorem 6.1 asserts the existence of nonconstant p-eriodic solutions for certain values of b_i, namely $b_i = b_i(p,\varepsilon) = \mu_i(p) + \gamma_i(p,\varepsilon)$ or a small interval of ε near $\varepsilon = 0$. If this map $(p,\varepsilon) \to (b_1,b_2)$ is one-one hen for any set of system parameters b_i there would be only one p-periodic solu-ion of a certain "amplitude" ε (at least only one on the bifurcating branch of heorem 6.1). However, if this map is not one-one, then for a set of values of b_i ore than one p-periodic solution will exist. We have not investigated this henomenon any further than this.

6.4 Oscillations and Predator-Prey Models with Delays. In his book Volterra 1931) (also see Volterra (1927)) showed that three "laws" concerning predator-prey nteractions which he had derived from the nondelay Volterra-Lotka model (4.1) were lso derivable from his delay model for a predator-prey interaction (4.2). These hree laws are the following:

Law of Oscillations. Both predator and prey densities oscillate indefinitely bout their respective equilibrium values.

Law of Conserved Means. The long time average of both prey and predator den-ities exists and equals their respective equilibrium values.

Law of Perturbed Means. If prey and/or predators are harvested proportionally to their densities, then the prey and/or predator equilibrium value increases and/or decreases respectively.

By "oscillate about equilibrium" e_1 is meant the following: $N_i(t) = e_i$ for an infinite sequence $t = t_n \to +\infty$ and $N_i(t)$ assumes infinitely many local maxima (greater than e_i) and minima (less than e_i). "Long time average equals e_i" means $\lim_{t \to \infty} t^{-1} \int_0^t N_i(s)ds = e_i$.

Volterra established these laws on the basis of the delay model

$$N_1'/N_1 = b_1 - d_1N_2 - a_{12} \int_{-\infty}^{t} N_2(s)k_{12}(t - s)ds$$

$$N_2'/N_2 = -b_2 + d_2N_1 + a_{21} \int_{-\infty}^{t} N_1(s)k_{21}(t - s)ds$$

$$k_{ij} \varepsilon L_+^1, \quad |k_{ij}|_1 = 1, \quad d_i \geq 0, \quad a_{ij} \geq 0, \quad d_i + a_{ij} \neq 0$$

under the assumption that $k_{ij}(t)$ has compact support: $k_{ij}(t) \equiv 0$ for $t \geq T > 0$ for some $T > 0$. In the case of the Law of Conserved Means it was necessary also to assume $d_1 > 0$ (i.e. that some instantaneous prey growth rate response is present). Volterra's proofs can also be carried out for models with Stieltjes integrals and hence for models with instantaneous lags. Since the details of Volterra' proof are lengthy and can be found in readily available references (e.g. Rescigno and Richardson (1973)) we will not give them here.

Volterra apparently did not study the question of the convergence or divergenc of the oscillations of his delay predator-prey model. We have seen in Chapter 4 that generally speaking the equilibrium in delay predator-prey models becomes unstable if the delay T in the system is long compared to the inherent growth rates b_i of both species or if the inherent prey carrying capacity c is large. With respect to long delays one can obtain first order approximations to the oscillaions, as was done in Section 5.1 for single species models, by doing a singular perturbation analysis of a general predator-prey system using the small dimensionless parameters $1/b_i T$. One example of this is given by Cushing (1977d) where the first order approximations show, as expected, divergent oscillations.

6.5 Two Species Competition Models with Linear Response Functionals. The most general two species competition model in which the per unit growth rates are linear functionals of species densities is

(6.29) $$N_i'/N_i = b_i\Big(1 - c_{ii}\int_{-\infty}^{t} N_i(s)d_s h_{ii}(t-s) - c_{ij}\int_{-\infty}^{t} N_j(s)d_s h_{ij}(t-s)\Big),$$

$$1 \le i \ne j \le 2, \quad b_i > 0, \quad c_{ii} \ge 0, \quad c_{ij} > 0, \quad \int_0^\infty dh_{ij}(s) = 1.$$

We assume that the equilibrium

(6.30) $$e_i = (c_{ii} - c_{ij})/\Delta, \quad j \ne i, \quad \Delta := c_{11}c_{22} - c_{12}c_{21}$$

is positive and isolated:

(6.31) $$\Delta \ne 0 \text{ and } e_i > 0.$$

If we attempt to apply the bifurcation Theorem 6.1 to this model, we must first consider the linearized system. The homogeneous algebraic system (6.5) for the mth Fourier coefficient of p-periodic solutions of this linearization has a coefficient matrix whose determinant is singular if and only if the two equations

a) $$m^2\xi_1\xi_2 - e_2c_{22}S_{22}(m)m\xi_1 - e_1c_{11}S_{11}(m)m\xi_2 + e_1e_2(c_{11}c_{22}\Sigma_1(m) + c_{12}c_{21}\Sigma_3(m)) = 0$$

(6.32)

b) $$e_2c_{22}C_{22}(m)m\xi_1 + e_1c_{11}C_{11}(m)m\xi_2 = e_1e_2(c_{11}c_{22}\Sigma_2(m) - c_{12}c_{21}\Sigma_4(m))$$

are satisfied for $\xi_i = \omega/\mu_i$ where

$$\Sigma_1(m) := S_{11}(m)S_{22}(m) - C_{11}(m)C_{22}(m)$$

$$\Sigma_2(m) := C_{11}(m)S_{22}(m) + C_{22}(m)S_{11}(m)$$

$$\Sigma_3(m) := C_{12}(m)C_{21}(m) - S_{12}(m)S_{21}(m)$$

$$\Sigma_4(m) := C_{12}(m)S_{21}(m) + C_{21}(m)S_{12}(m)$$

$$\int_0^\infty e^{-im\omega s} dh_{ij}(s) := C_{ij}(m) - iS_{ij}(m).$$

These equations (6.32) must be solvable for $m = 1$ (without loss in generality, see Theorem 6.2) for positive $\xi_i = \omega/\mu_i$ as is necessary for p-periodic bifurcation $p = 2\pi/\omega$ at the critical values $\mu_i > 0$ of the inherent growth rates $b_i > 0$. In addition these equations must not be satisfied for these same roots ξ_i for $m \geq 2$ in order to guarantee the sufficiency condition that $2 = n \geq r = 2$ in Theorem 6.1. Finally, the sufficiency condition $\det C^* \neq 0$ must also hold.

To give general conditions on the S_{ij} and C_{ij} (i.e. on the delay integrators and the period p) under which all of these conditions hold and hence bifurcation occurs is complicated. First of all, the necessary condition that (6.32) be solvable for $m = 1$ for positive roots $\xi_i > 0$ requires that the hyperbola (6.32a) intersect the straight line (6.32b) in the first quadrant $\xi_i > 0$ for $m = 1$. Although it is not difficult to state conditions on the $S_{ij}(1)$, $C_{ij}(1)$ for which this is true, the sufficiency conditions (that this point not be an intersection point when $m > 1$ and that $\det C^* \neq 0$) are very complicated to relate directly to $S_{ij}(1)$, $C_{ij}(1)$. Thus, we will consider in this general setting only the special case when no delays are present in the self-inhibitation terms.

Suppose $h_{ii}(s) = u_0(s)$ and at least one $c_{ii} \geq 0$, say c_{11}, is nonzero. Thus, we consider the model

$$(6.33) \qquad \begin{gathered} N_i'/N_i = b_i\left(1 - c_{ii}N_i - c_{ij}\int_{-\infty}^{t} N_j(s)d_s h_{ij}(t-s)\right) \\ c_{11} > 0, \quad c_{22} \geq 0, \quad \int_0^\infty dh_{ij}(s) = 1, \quad c_{ij} > 0, \quad i \neq j \end{gathered}$$

under the assumption (6.31). In this case, when delays are present only in the interspecies interaction terms, the equations (6.32) reduce to

$$(6.34) \qquad \begin{aligned} &\text{(a)} \qquad m^2\xi_1\xi_2 + e_1e_2(-c_{11}c_{22} + c_{12}c_{21}\Sigma_3(m)) = 0 \\ &\text{(b)} \qquad e_2c_{22}m\xi_1 + e_1c_{11}m\xi_2 = -e_1e_2c_{12}c_{21}\Sigma_4(m) \end{aligned}$$

which are solvable for $\xi_i > 0$ and $m = 1$ only if

$$(6.35) \qquad \Sigma_4(1) < 0 \quad \text{and} \quad c_{12}c_{21}\Sigma_3(1) < c_{11}c_{22}.$$

If (6.34a) when $m = 1$ is solved for ξ_1 in terms of ξ_2 and the result is substituted into (6.34b) when $m = 1$, one finds that ξ_2 must solve the quadratic

$$(6.36) \qquad c_{11}\xi_2^2 + e_2c_{12}c_{21}\Sigma_4(1)\xi_2 + e_2^2c_{22}(c_{11}c_{22} - c_{12}c_{21}\Sigma_3(1)) = 0.$$

Thus, we require that the discriminant of this quadratic be positive

$$(6.37) \qquad c_{12}^2c_{21}^2\Sigma_4^2(1) > 4c_{11}c_{22}(c_{11}c_{22} - c_{12}c_{21}\Sigma_3(1))$$

in which case there are either two distinct roots $\xi_2 > 0$ (when $c_{22} \neq 0$) or one root $\xi_2 > 0$ (when $c_{22} = 0$). The value of ξ_1 is then determined from either equation in (6.34).

Thus (6.35) and (6.37), as conditions on the coefficients c_{ij}, the delay integrators in the interspecies interaction terms of (6.33) and on the period p, are necessary for bifurcation of nontrivial p-periodic solutions from equilibrium.

The condition that ξ_1, ξ_2 not satisfy (6.34) for $m > 1$ is clearly met if for each $m > 1$

$$(6.38)\quad \begin{cases} \text{either} \quad (a) \quad \Sigma_4(m) \neq m\Sigma_4(1) \\ \text{or} \quad (b) \quad c_{12}c_{21}\Sigma_3(m) - c_{11}c_{22} \neq m^2(c_{12}c_{21}\Sigma_3(1) - c_{11}c_{22}). \end{cases}$$

Finally, the determinant $\det C^*$ must be nonzero. The linearized system has two independent p-period solutions y^1 and y^2 given by

$$y^1 = \text{col}\,(-\mu_1 e_1 c_{12}(C_{12} \cos \omega t + S_{12} \sin \omega t), \quad \mu_1 e_1 c_{11} \cos \omega t - \omega \sin \omega t)$$

$$y^2 = \text{col}\,(-\mu_1 e_1 c_{12}(-S_{12} \cos \omega t + C_{12} \sin \omega t), \quad \omega \cos \omega t + \mu_1 e_1 c_{11} \sin \omega t)$$

where $\mu_i = \omega/\xi_i$. The adjoint linear system has two independent p-periodic solutions

$$z^1 = \text{col}\,(\mu_2 e_2 c_{22} \cos \omega t + \omega \sin \omega t, -\mu_1 e_1 c_{12}(C_{12} \cos \omega t - S_{12} \sin \omega t))$$

$$z^2 = \text{col}\,(-\omega\cos \omega t + \mu_2 e_2 c_{22} \sin \omega t, -\mu_1 e_1 c_{12}(S_{12} \cos \omega t - C_{12} \sin \omega t)).$$

If $y = \kappa_1 y^1 + \kappa_2 y^2$ and $\det C^*$ is computed it turns out that

$$\det C^* = \frac{1}{4}(S_{12}^2 + C_{12}^2)(\mu_1 e_1 c_{12})^2(\mu_1 e_1 c_{11} - \mu_2 e_2 c_{22})(\kappa_1^2 + \kappa_2^2).$$

nce $\Sigma_4(1) < 0$ (see (6.35)) it follows that $S_{12}^2 + C_{12}^2 \neq 0$ and that $\det C^*$ if d only if $\mu_1 e_1 c_{11} \neq \mu_2 e_2 c_{22}$. This latter condition is equivalent to $e_1 c_{11} \xi_2 \neq$ $c_{22}\xi_1$ which is easily shown to hold by recalling that the ξ_i satisfy (6.34) d that (6.37) holds.

Theorem 6.6 _If the interspecies delay integrators_ $h_{ij}(s)$ _in the general two_ _ecies competition model_ (6.33) _under the assumption_ (6.31) _satisfy_ (6.35), (6.37) _d_ (6.38) _for some period_ $p = \omega/2\pi$, _then nontrivial_ p-_periodic solutions bifur-_ _te from the positive equilibrium_ (6.30) _for inherent growth rates_ b_i _near the_ _itical values_ $\mu_i = \omega/\xi_i$ _where_ $\xi_2 > 0$ _is a positive root of the quadratic_ .36) _and_

$$\xi_1 = e_1 e_2 (c_{11} c_{22} - c_{12} c_{21} \Sigma_3(1))/\xi_2.$$

Note that (6.37) automatically holds if $c_{22} = 0$ in which case (6.35) reduces $\Sigma_4(1) < 0$, $\Sigma_3(1) < 0$.

Also note that the hypotheses (6.35), (6.37) and (6.38) of Theorem 6.6 are of o types. Condition (6.38a) and the first inequality of (6.35) involve only the lay integrators and the period p. Only the remaining condition (6.37) and the cond inequality in (6.35) involve the interaction coefficients c_{ij}. With regard these latter two inequalities we observe that they may be rewritten as

$$\lambda > \Sigma_3(1), \quad p(\lambda) := 4\lambda^2 - 4\Sigma_3(1)\lambda - \Sigma_4^2(1) < 0$$

ere $\lambda > 0$ is the ratio $\lambda = c_{11}c_{22}/c_{12}c_{21}$. These inequalities hold if and only

(6.39) $$\Sigma_3(1) < \lambda_0$$

and

(6.40) $$\Sigma_3(1) < \lambda < \lambda_0,$$

where $\lambda_0 > 0$ is the unique positive root of the quadratic $p(\lambda)$, i.e.

$$\lambda_0 = \frac{1}{2}\left(\Sigma_3(1) + \sqrt{\Sigma_3^2(1) + \Sigma_4^2(1)}\right).$$

It is easy to see that (6.39) in fact holds.

COROLLARY 6.7 <u>The conclusions of Theorem</u> 6.6 <u>hold for the competition model</u> (6.33) (<u>under</u> (6.31)) <u>if the delay integrators and the period</u> p <u>satisfy</u> $\Sigma_4(1) < 0$ <u>and</u> (6.38a) <u>and if the ratio</u> $\lambda = c_{11}c_{22}/c_{12}c_{21}$ <u>satisfies</u> (6.40).

The ratio λ measures the relative "strengths" of the intraspecies interaction (measured by the product $c_{11}c_{22}$) and the interspecies interaction (measured by $c_{12}c_{21}$). If $\lambda > 1$ then $\Delta > 0$ and (6.31) implies $c_{ii} > c_{ij}$ while if $\lambda < 1$ then $c_{ii} < c_{ij}$. It is well known (see J. M. Smith (1974), Chapter 5) that <u>when no delays are present</u> in (6.33), $\lambda > 1$ implies that the equilibrium is (globally) A.S. while if $\lambda < 1$ this equilibrium is unstable, each solution (N_1, N_2) tending (in the limit as $t \to +\infty$) to either $E_1 := (0, 1/c_{22})$ or $E_2 := (1/c_{11}, 0)$ depending on the initial state. The condition $\lambda > 1$ can be said to be the stable case when intraspecies competition is stronger than interspecies competition while the case $\lambda < 1$ is the unstable case of when interspecies competition is stronger than intraspecies competition. Note that when no delays are present ($S_{ij} = 0$, $C_{ij} = 1$) the condition $\Sigma_4(1) < 0$ necessary for bifurcation fails to hold (since in this case $\Sigma_4(1) = 0$). Thus, nontrivial periodic oscilla-

ons bifurcation from equilibrium only if delays are present in the general compe-
tion model (6.33).

It is of interest to note that it is possible for $\lambda < 1$ in Corollary 6.7 and a result for nontrivial periodic solutions to exist in the case when interspecies npetition is the stronger. This suggests that the notion of "ecological niche" "competitive exclusion", as based on the instability of the positive equilibrium d the attractivity of the equilibria E_1 and E_2 for the nondelay model in the se of strong interspecies competition, may be affected by delays in growth rate sponses to interspecies contacts in the sense that the species could coexist in undamped oscillatory manner.

Example (1) Suppose both delays in (6.33) are "weak" generic delays
$_{ij}(s) = T_i^{-1} \exp(-t/T)$, $T_i > 0$. Then

$$S_{ij}(1) = \omega T_i(1 + \omega^2 T_i^2)^{-1}, \qquad C_{ij}(1) = (1 + \omega^2 T_i^2)^{-1}$$

$$\Sigma_4(1) = (T_1 + T_2)(1 + \omega^2 T_1^2)^{-1}(1 + \omega^2 T_2^2)^{-1} > 0.$$

ce the necessary condition $\Sigma_4(1) < 0$ fails to hold there is no bifurcation in s case.

Example (2) Suppose both delays in (6.33) are "strong" generic delays
$_j(s) = T^{-2} t \exp(-t/T)$ of equal delay measure $T > 0$. Then

$$S_{ij}(m) = 2m\omega T(1 + m^2\omega^2 T^2)^{-2}$$

$$C_{ij}(m) = (1 - m^2\omega^2 T^2)(1 + m^2\omega^2 T^2)^{-2}$$

and

$$\Sigma_4(m) = 4m\omega T(1 - m^2\omega^2T^2)(1 + m^2\omega^2T^2)^{-4}.$$

The necessary condition $\Sigma_4(1) < 1$ holds if and only if

$$\omega^2T^2 > 1 \quad \text{or} \quad p < 2\pi T. \tag{6.41}$$

It is not difficult to show that $\Sigma_4(m) \neq m\Sigma_4(1)$ for all $m \geq 2$ and hence (6.38a) holds. Finally, it turns out that

$$\lambda_0 = (1 - \omega^2T^2)^2(1 + \omega^2T^2)^{-4} < 1$$

so that (6.40) holds and bifurcation occurs (by Corollary 6.7) provided the coefficients satisfy

$$\lambda = c_{11}c_{22}/c_{12}c_{21} < (1 - \omega^2T^2)^2(1 + \omega^2T^2)^{-4}. \tag{6.42}$$

Note that $\lambda < 1$ so that this result applies only in the case of strong interspecies interaction.

Given c_{ij} such that $\lambda < 1$ and a delay T can a period p be found such that (6.41) and (6.42) hold? The maximum value that the right hand side of the inequality (6.42) can assume is $1/64$; which occurs when $\omega^2T^2 = 3$ or $p = 2\pi T/\sqrt{3}$. In any case the given coefficients must satisfy $\lambda < 1/64$, i.e. $64c_{11}c_{22} < c_{12}c_{21}$ which means that interspecies competition must be relatively strong. □

Although the possible coexistence of two competing species under strong interspecies competition is hinted at by the existence of nonconstant periodic solutions in Theorem 6.6 above, this in itself is not necessarily a strong statement in vor of this possibility in view of the lack of any assertion concerning the ability of these periodic solutions. The positive equilibrium (6.30) is unstable der strong interspecies competition for all delay kernels, interaction coeffi-ents c_{ij} and inherent birth rates b_i (Theorem 4.13). Thus, it is not pos-ble that there is an "exchange of stability" from this equilibrium to a bifurcat-g limit cycle as is often the case for differential equations. It seems unlikely at the period solutions of Theorem 6.6 are stable and this is born out by the thor's numerical integrations of the model in Example (2) in which no limit cycle havior was found. On the other hand, there could be a stable manifold of solu-ons associated with the periodic solutions, a possibility we have not investi-ted.

Also relative to this point is the stability of the equilibria E_1 and E_2 the delay model. If (6.29) is linearized about one of these equilibria, the re-lting linear system has characteristic equation

$$[z - b_j(c_{ii} - c_{ji})c_{ii}^{-1}][z + b_i e_i c_{ii} k_{ii}^*(z)] = 0.$$

us, if interspecies competition is strong $c_{ji} > c_{ii}$ and each species in the sence of the other behaves according to an A.S. delayed logistic (so that $+ b_i k_{ii}^*(z) = 0$ has no roots $\operatorname{Re} z \geq 0$), then each of these equilibria is locally S. Thus (as in the nondelay case) one species will go extinct, at least locally en the populations initially are near one of these equilibria.

6.6 Two Species Mutualism Models with Linear Response Functionals. The system

$$N_i'/N_i = b_i(-1 - c_{ii}\int_{-\infty}^{t} N_i(s)d_s h_{ii}(t-s) + c_{ij}\int_{-\infty}^{t} N_j(s)d_s h_{ij}(t-s)),$$

.43)

$$1 \leq i \neq j \leq 2, \quad b_i > 0, \quad c_{ii} \geq 0, \quad c_{ij} > 0, \quad \int_0^\infty dh_{ij}(s) = 1$$

is the most general model, with a linear growth rate response functional, in which both species die in the absence of the other while the interaction of both is mutually beneficial. Thus, the mutualism here is obligate to both species. Although mutualistic interactions are not found as frequently in nature as competitive or predator-prey interactions, they occur often enough to be of more than passing interest to ecologists (Trager (1970), Ricklefs (1973)). Nonetheless there seem to be no differential systems which have been offered as models for mutualistic interactions. The simplest nondelay models, namely those with linear responses ($h_{ij} = u_0$ in (6.43)), have the undesirable property that either both species densities tend to zero or to $+\infty$ as $t \to +\infty$. May (1974, p. 224) points out that many mutualistic interactions (for example, between plants and pollinators) characteristically involve significant delays in growth rate response.

If the model (6.43), under the assumption that there exists a positive equilibrium,

$$(6.44) \qquad e_i = -(c_{jj} + c_{ij})/\Delta, \qquad \Delta := c_{11}c_{22} - c_{12}c_{21} < 0$$

is investigated by means of Theorem 6.1 for the possible bifurcation of nonconstant periodic solutions, one finds that the details are exactly the same as those in the previous Section 6.1 for competition models with linear response functionals with only one exception: one of the components of the linearized periodic solution y and also the adjoint solutions Z^1 and Z^2 change sign. Thus, <u>with the added assumption that $\Delta < 0$, Theorem 6.6 and Corollary 6.7 remain valid as stated, but for the mutualism model</u> (6.43). (The details of the two Examples (1) and (2) also apply to (6.43) as illustrations.) This hints at the possible coexistence of the two species, for appropriate inherent death rates b_i, in a mutualistic interaction when delays are present in growth rate responses. However, the remarks concerning stability at the end of the previous Section 6.5 also apply here. Numerical inte-

:ations carried out by the author failed to find any limit cycle behavior.

The positive equilibrium (6.44) (Theorem 4.13) and the origin $N_1 = N_2 = 0$ e unstable for all coefficients d_{ij}, death rates b_i and delay integrators $_j$. There are no other equilibria in the first quadrant or on the N_1 or N_2 :is.

6.7 Delays in Systems with More than Two Interacting Species. As the number species increases the complexity of the model and the possible behavior of its lutions and of course the mathematical analysis needed to study the model also eatly increase. With regard to the possible existence of nonconstant periodic lutions as given by Theorem 6.1, the algebraic details become formidable as the ze of the system increases. We will not attempt to make any general applica- ons of Theorem 6.1, but will restrict our remarks to some specific three species dels.

While the two species interaction categories of predator-prey, competition, tualism, etc. serve as elementary building blocks in the ecological study of mmunities of species, ecologists recognize that two species rarely, if ever, teract solely with each other in the absence of other species. In an attempt to tter understand some of the fundamental concepts in ecology as they concern more an two interacting species, a few specific three species models have been stud- d in the literature. For example, two predator-one prey models have been stud- d with respect to competition and the fundamental law of exclusion or "niche" och (1974), Caswell (1972)). Two competing prey-one predator models have also en investigated to see what effect predation has on competition (Parrish and ila (1970), Cramer and May (1971)). Systems of three competing species have been vestigated (e.g. see MacArthur and Levins (1967), May and Leonard (1975), scigno (1968)).

The only three species models with delays in the growth rate responses which seem to have been considered in the literature are two predator-one prey models in Caswell (1972) and Cushing (1977c) and a mutualism model in Cushing (1977c). The numerical simulations done by Caswell show the nature of the effect on oscillations caused by delays and the possible reversal of the outcome of the competition between the two predators which can be caused by the presence of delays. The two models considered by Cushing (1977c) serve to illustrate the bifurcation Theorem 6.1 as applied to certain three species models.

In order to illustrate some of the features of the bifurcation Theorem 6.1 as applied to a three species model we will very briefly consider a two prey-one predator model (with linear response functionals) considered by Parrish and Saila (1970) and Cramer and May (1972), but (as in Volterra's original delay predator-prey model) with delays in the predator's response to prey density changes:

$$(6.45)\quad \begin{aligned} N_1'/N_1 &= b_1(1 - c_{11}N_1 - c_{12}N_2 - c_{13}N_3) \\ N_2'/N_2 &= b_2(1 - c_{21}N_1 - c_{22}N_2 - c_{23}N_3) \\ N_3'/N_3 &= b_3\left(-1 + c_{31}\int_{-\infty}^{t} N_1(s)d_sh_1(t-s) + c_{32}\int_{-\infty}^{t} N_2(s)d_sh_2(t-s)\right) \\ & b_i > 0, \quad c_{ij} > 0, \quad \int_0^{\infty} dh_i(s) = 1. \end{aligned}$$

Here N_1 and N_2 are the densities of two prey species which in the absence of predator N_3 are in competition for a common resource. The only delays are in the (linear) response of the predator to prey density changes. The papers by Parrish and Saila and Cramer and May study the possible stability of a positive equilibrium for the nondelay version of this model $(h_i(s) = u_0(s))$ under the assumption that the competing species N_1 and N_2 are unstable in the absence of the predator. First of all, one must assume that the coefficients c_{ij} are such that a positive equilibrium exists; this is straightforward, but rather tedious to rite down (see Parrish and Saila (1970), equations (14) - (16)). Secondly, the igenvalues of the linearized system must be computed. The question then becomes: an these eigenvalues all lie in the left half plane (so that (6.45) has an A.S. quilibrium) under the added assumption that $c_{11}c_{22} < c_{12}c_{21}$ (which is equivalent the instability of the prey competition in the lack of predation)? Cramer and y show that this is possible.

If in order to see if delays can lead to periodic oscillations we wish to apy the bifurcation theory of Section 6.1, then we are led to the following two quations for the three critical growth rates μ_i:

$$\beta_{22}\xi_1\xi_3 + \beta_{11}\xi_2\xi_3 - \beta_{23}\beta_{32}S_2\xi_1 - \beta_{13}\beta_{31}S_1\xi_2$$

$$= \beta_{32}(\beta_{11}\beta_{23} - \beta_{21}\beta_{13})C_2 + \beta_{31}(\beta_{22}\beta_{13} - \beta_{12}\beta_{23})C_1$$

$$\xi_1\xi_2\xi_3 - \beta_{23}\beta_{32}C_2\xi_1 - \beta_{13}\beta_{31}C_1\xi_2 + (\beta_{12}\beta_{21} - \beta_{11}\beta_{22})\xi_3$$

$$= -\beta_{32}(\beta_{11}\beta_{23} - \beta_{21}\beta_{13})S_2 - \beta_{31}(\beta_{22}\beta_{13} - \beta_{12}\beta_{23})S_1$$

$$\xi_i = \omega/\mu_i, \qquad \omega = 2\pi/p, \qquad \beta_{ij} = e_i c_{ij}$$

$$S_i = \int_0^\infty \sin \omega s dh_i(s), \qquad C_i = \int_0^\infty \cos \omega s dh_i(s).$$

e reason for our writing these equations is to demonstrate the type of equations be solved for the critical values μ_i in typical applications of Theorem 6.1 r $n \geq 3$ species. It is possible to place (inequality type) conditions on S_i d C_i (i.e. on the period p) so that these equations may be solved for ξ_1, in terms of ξ_3 and hence the critical value of the predator's death rate μ_3 arbitrary (or at least constrained to some interval). Moreover, the matrix C* hose rank must be made equal to 2) is 2×3 and hence, as discussed in Remark

(2) of Section 6.1, the higher order term in b_3 is also arbitrary. In other words, it is possible to find a bifurcating branch of p-periodic solutions with respect to the parameters b_1, b_2 for arbitrary b_3 and p (in certain intervals).

6.8 Periodically Fluctuating Environments. We have dealt in this chapter with periodic oscillations caused by delays in growth rate responses to density changes. Oscillations in species densities can also be a result of periodic change in environmental parameters. Mathematically, the responses functionals f_i which describe the per unit growth rates in a model would, in such a case, depend explicitly on time t. If this dependence is p-periodic, it is natural to ask for conditions under which the resulting p-periodic, nonautonomous system would have a (hopefully stable) p-periodic solution. This problem was discussed in Section 5.4 for the single, isolated species case. When no delays are present this problem was studied by Cushing (1976b) for n-species communities and by Cushing (1977e) for predator-prey interactions. The approach in these papers was to show the existence of a positive, p-periodic by proving that a bifurcation of such solutions occurs as the average of the (now p-periodic time dependent) inherent growth rate of one of the species, say the nth, passes through a critical value. The bifurcation occurs from a positive, p-periodic solution of the (time dependent) subcommunity obtained by eliminating the nth species. Starting then with the results for one species (Section 5.4) one can then derive, by a repeated application of the theorems in Cushing (1976b, 1977e), the existence of positive, p-periodic solutions of time dependent n-species models for appropriate values of the averages of the inherent growth rates.

Although the details of these assertions are carried out in the abovementioned references only for nondelay models with linear response functionals, the arguments can be carried over almost verbatim to more general delay models with possibly non-

:near response functions. This is in fact explicitly pointed out by Cushing .977e) for predator-prey models.

CHAPTER 7. SOME MISCELLANEOUS TOPICS

These notes have dealt exclusively with the stability or instability of equilibria or with the nature of oscillations of solutions of ecological models which incorporate delays in the growth rate responses of the member species of a community. There are of course a great many other topics which could also be of interest. Any question one might ask concerning the dynamics of a community, one can ask of the functional models of the type we have been considering in order to see what effects, if any, response delays would have on the answer.

For example, if an element of randomness is assumed in the environment one is then confronted with a random differential equation or, if delays are present, a random integrodifferential equation. Although many such nondelay models have been studied (e.g. see Lin and Kahn (1977), May (1974) and their cited references) and although some random models with time lags have been considered (e.g. see White (1977)), random integrodifferential models in ecology seem not to have been studied in the literature.

Another essentially unexplored topic is the effect that delays have on a community of species which, besides evolving in time, are allowed to diffuse spacially. Such a situation can be modeled by adding a diffusion term to the functional equations:

$$\partial N_i/\partial t = d_i \Delta N_i + N_i f_i, \qquad \Delta N_i = \Sigma \partial^2 N_i/\partial x_i^2$$

where the diffusion constant d_i is positive and the x_i are space variables. If no delays are present $f_i = f_i(N_1, \ldots, N_n)$, then this (nonlinear) parabolic system is of a well-known, well-studied type. Such so-called reaction-diffusion equations have been studied in a variety of biological and chemical contexts; they have recently been appearing frequently in the literature as models in ecological contexts as well. The effect of time delays in the response functional f_i has, to e author's knowledge, been virtually unexplored. The only papers known to the thor are those of Murray (1976), Dunkel (1968a), Wang (1963), Wang and Bandy 963), and Scott (1969), all of which deal with systems in which there appear con- ant time lags. Dunkel (1968a) is concerned only with the existence of solutions a single linear equation. Murray (1976) considers the interesting question of able, traveling waves solutions, which he proves exist for certain (single) non- near equations. Thus, Murray suggests that spacial and time fluctuations can be used by time delays, as separate from the reaction-diffusion mechanism which, as well known, can be a source of such spacial inhomogeneities in and of itself. ng et al. (1963) consider the stability of equilibria using Liapunov methods for rabolic systems with constant time lags. Scott (1969) considers the extent to ich the method of separation of variables can be applied to linear systems when nstant time lags are present. Only Murray's paper is related specifically to ological matters and deals with nonlinear equations. No papers appear in the terature which deal with delay models of the more general Volterra integral form for delay systems of more than one equation.

Another relatively unexplored question is that concerning the effect of delays forced (or controlled) models. If some mechanism affects the rate of change of e species densities independently from the size of the densities, then the models the type we have been considering here become nonhomogeneous:

$$N_i' = N_i f_i + h_i(t), \qquad 1 \leq i \leq n.$$

re $h_i(t)$ describes the state-independent rate of change of N_i which might sult, example, from harvesting and/or seeding, immigration and/or emigration, c. Although this model has been investigated for a variety of specific ecological teractions in the absence of delays, little has been done when delays are present. olle (1976) gives conditions under which a delay predator-prey model of the

Cunningham-Wangersky type (Section 4.5) with a positive, periodic forcing function in the predator equation has a periodic solution. This question is of interest in, amongst other things, immunology where the model describes the dynamics between a bacteria (prey) and an antibody (predator) when periodic injections of antibody are made. Cushing (1976d) gives conditions under which solutions of a forced, delay Volterra predator-prey model are asymptotically periodic. Brauer (1977a,b) studies the stability of equilibria of harvested, single species models with delays. This, to the author's knowledge, is the extent of the literature on forced, integrodifferential models in population dynamics.

As an example of the effect that delays can have on a forced equation consider the case of the constant rate harvesting of a single species whose unharvested density obeys a logistic law:

$$N' = bN(1 - c^{-1}N) - H \tag{7.1}$$

$$b > 0, \quad c > 0, \quad H > 0.$$

If the species has a delay in its growth rate response, then the model becomes

$$N' = bN\left(1 - c^{-1}\int_{-\infty}^{t} N(s)k(t - s)ds\right) - H \tag{7.2}$$

$$\int_{0}^{\infty} k(s)ds = 1.$$

The less realistic (and generally more difficult to analyze) model with a constant time lag is considered by Brauer (1977a) as well as models with a more general response functional. The following facts are elementary to show as far as the nondelay model (7.1) is concerned:

(a) if $H \varepsilon (0,H_0)$ where $H_0 = bc/4$ then (7.1) has two positive equilibria

$e^{\pm}$: $0 < e^- < e^+ < c$ the larger of which e^+ is A.S. and the smaller of which e^- is unstable;

(b) if $H > H_0$ then (7.1) has no equilibria and every solution (initially positive) vanishes in finite time.

us, H_0 is a critical harvesting load below which constant rate harvesting can done with a stable population density as a result, but above which harvesting ads to extinction in finite time. We ask: to what extent are these conclusions ue for the delay model (7.2)?

First of all, concerning (b) we have the following weaker, but qualitatively nilar result for the delay model (7.2).

THEOREM 7.1 Suppose $H > H_0$ and $k(t) \geq 0$, $t \geq 0$ in (7.2) and suppose t) is a solution with positive, bounded initial values $(t \leq 0)$. Either $N(t)$ nishes in finite time or $\liminf_{t \to \infty} N(t) = 0$.

is result, while weaker than (b), is enough for us to conclude that for large nstant rate harvesting $H > H_0$ in the delay model (7.2) the population will go tinct.

Proof. First, observe that $H > bc/4$ implies $be(1 - c^{-1}e) - H < 0$ for all and consequently that (7.2) has no equilibrium. Secondly, if $N(t)$ is a solu- on with bounded initial values then $N(+\infty) = \lim_{t \to +\infty} N(t)$ cannot be finite. is can be deduced as follows. If $N(+\infty)$ is finite, then $N(t)$ is bounded for $t \varepsilon (-\infty, +\infty)$ from which (together with (7.2)) follows the boundedness of $N'(t)$ hence the uniform continuity of $N(t)$ for all t. Thus, from (7.2) the deriva- e $N'(t)$ is uniformly continuous for all t and Barbalat's lemma (Barbalat 59)) implies that $N'(+\infty) = 0$. But then (7.2) implies $bN(+\infty)(1 - c^{-1}N(+\infty)) - H$, in contradiction to $H > bc/4$.

Suppose that $N(t)$ does not vanish in finite time. Then $N(t) > 0$ for all t.

Suppose for the purposes of contradiction that $\liminf_{t \to +\infty} N(t) = \delta > 0$. The fact that $H > bc/4$ implies that $b\delta(1 - c^{-1}\delta) - H < 0$. Let $\varepsilon > 0$ be a number so small that

$$(7.3) \qquad b\delta(1 - c^{-1}\delta) - H + b\delta c^{-1}\varepsilon < 0$$

and let t_0 be so large that $N(t) \geq \delta - \varepsilon$ for all $t \geq t_0$. Since $N(+\infty)$ does not exist, $N(t)$ oscillates and we may choose a sequence such that

$$t_n \geq t_0, \qquad t_n \to +\infty, \qquad N(t_n) \to \delta, \qquad N'(t_n) = 0.$$

Now

$$\int_{-\infty}^{t_n} N(s)k(t_n - s)ds \geq \int_{t_0}^{t_n} N(s)k(t_n - s)ds \geq (\delta - \varepsilon) \int_0^{t_n - t_0} k(s)ds.$$

Thus, from (7.2) at $t = t_n$

$$0 \leq bN(t_n)(1 - c^{-1}(\delta - \varepsilon) \int_0^{t_n - t_0} k(s)ds) - H$$

which implies, upon our letting $t_n \to +\infty$, that

$$0 \leq b\delta(1 - c^{-1}(\delta - \varepsilon)) - H.$$

This contradiction to (7.3) implies $\delta = 0$. □

If now we assume $H \in (0, H_0)$ then the delay model (7.2) has two (exactly the same two as (7.1)) equilibria:

$$e^{\pm} = e^{\pm}(H) = (bc \pm \sqrt{bc(bc - 4H)})/2b > 0.$$

is easy to show that, as in the nondelay case, the smaller equilibrium e^- is ays unstable.

THEOREM 7.2 _If_ $H\varepsilon(0,H_0)$, $H_0 = bc/4$, _then the equilibrium_ e^- _of_ (7.2) _is_ _table_.

Proof. If (7.2) is linearized about e^- the resulting linear integrodiffer- ial equation has characteristic function $p(z) = z - b(1 - e^-/c) + e^-bk^*(z)/c$. e that $p(0) = b(2e^-c^{-1} - 1) < 0$. If $z = x > 0$ is real then since $k^*(x)$ is ided, $p(+\infty) = +\infty$ and as a result $p(z)$ has a positive real root. □

The investigation of the larger equilibrium e^+ is more difficult. Suppose let $k(t)$ be the "strong" generic kernel $k(t) = T^{-2}t \exp(-t/T)$, $T > 0$. If 2) is linearized about e^+, the resulting linear equation has characteristic tion $p(z) = N(z)/(zT + 1)^2$ where

$$N(z): = T^2z^3 + T(bTc^{-1}e^+ + 2 - bT)z^2 + (2bTe^+c^{-1} + 1 - 2bT)z + b(2e^+c^{-1} - 1).$$

s, e^+ is A.S. if and only if all roots of N lie in the left half plane. The fficient $b(2e^+c^{-1} - 1)$, which is independent of the delay T, is easily shown e positive. Using the Hurwitz criteria one finds for $H\varepsilon(0,H_0)$ that e^+ is if and only if both of the following inequalities hold:

$$T < T_1(H): = c/2b(c - e^+)$$

$$2b^2(c - e^+)^2T^2 - bc(4c - 3e^+)T + 2c^2 > 0.$$

An investigation of the later inequality shows that it holds if and only if $T \notin (T_2(H), T_3(H))$ where T_2, T_3 are the two positive real roots of the quadratic expression on the left hand side. Here it is used that $e^+(H)$ is a monotonically decreasing function from c to $c/2$ for $H \varepsilon [0, H_0]$. Moreover, it can be shown that $T_2(H) < T_1(H) < T_3(H)$ for all $H \varepsilon (0, H_0)$ and that

$$(7.4) \qquad T_2(H) = [c(4c - 3e^+) - \sqrt{e^+(8c - 7e^+)}]/b(c - e^+)^2.$$

Thus, we have the following result.

THEOREM 7.3 <u>Suppose</u> $H \varepsilon (0, H_0)$, $H_0 = bc/4$, <u>and</u> $k(t) = T^{-2} t \exp(-t/T)$, $T > 0$ <u>in the model</u> (7.2). <u>There exists a continuous, monotonically decreasing function</u> $T_2(H)$ (<u>given by</u> (7.4)) <u>defined for</u> $H \varepsilon [0, H_0]$ <u>satisfying</u> $T_2(0) = 2/b$, $T_2(H_0) = 1/b$ <u>such that the positive equilibrium</u> e^+ <u>is (locally) A.S. for</u> $0 < T < T_2(H)$ <u>and unstable for</u> $T > T_2(H)$ <u>for each</u> $H \varepsilon (0, H_0)$.

FIGURE 7.1 graphically illustrates the stability properties of e^+ as a function of the parameters T and H.

Note that the effect of large delays $(T > 1/b)$ in a constant rate harvested logistic-growth species model (7.2) is to decrease the critical harvesting constant from H_0 to some value $H(T)$ less than H_0 given by the inverse of the function (7.4) for $T < 2/b$ or $H(T) = 0$ for $T \geq 2/b$. For large delays $T > 2/b$ no amount of constant rate harvesting will result in a stable equilibrium. For small delays $T < 1/b$ the critical harvesting constant H_0 at which instability occurs is the same as that of the nondelay model. The fact that for a given $H \varepsilon (0, H_0)$ the equilibrium e^+ loses its stability as the delay T passes a critical value $T_2(H)$ suggests the possibility of the bifurcation of nonconstant periodic solutions at this value of T. This phenomenon can be investigated for the "strong"

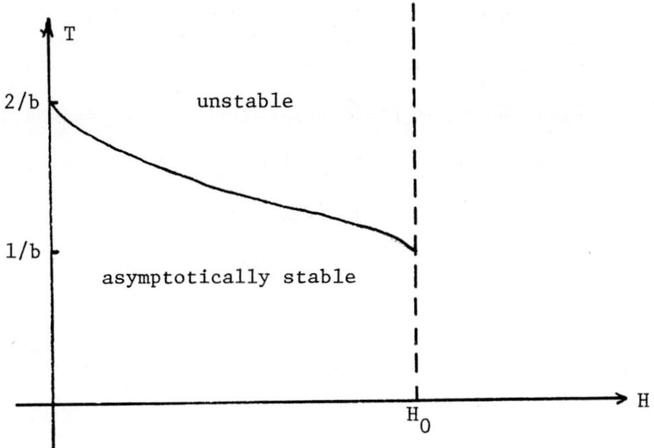

The H and T regions of stability and instability for the larger equilibrium e^+ of the delay logistic, constant rate harvested model (7.2) for fixed birth rate b and inherent carrying capacity c.

FIGURE 7.1

neric delay kernel by means of the Hopf bifurcation theorem in a manner similar that used in Section 5.2 where the unharvested case $H = 0$ was considered. us, for this kernel (7.2) is equivalent, as far as periodic solutions are conrned, to the nondelay system

$$x_1' = b(1 - e^+c^{-1})x_1 - be^+c^{-1}x_2 - bc^{-1}x_1x_2$$

$$x_2' = x_3$$

$$x_3' = T^{-2}(x_1 - x_2) - 2T^{-1}x_3$$

ere $x_1 = N_1 - e^+$ and $x_2 = \int_{-\infty}^{t} x_1(s)k(t - s)ds$. Since the algebraic details e quite formidable in the general case, we carried out the Hopf bifurcation analis of this system for $b = 1$ and that value of H for which $e^+c^{-1} = 3/4$ (i.e. at harvesting constant for which the equilibrium e^+ is 75% of the inherent carrying capacity c). The critical value of the delay turns out to be (from (7.4)) $T_0 = 7 - \sqrt{33} \sim 1.26$. For this value of T the characteristic function or what amounts to the same thing the cubic polynomial $N(z)$ has a conjugate pair of purely imaginary roots $z = \pm iy \sim \pm 0.49i$. If $z = z(T)$ denotes the branch of roots of $N(z)$ which at $T = T_0$ equals iy, then an implicit differentiation of $N(z(T)) = 0$ yields $z'(T)$ and from this we have calculated $\text{Re } z'(T_0) \sim 0.85 > 0$. The fact that $\text{Re } z'(T_0) > 0$ insures that Hopf bifurcation occurs. In order to determine the characteristics of the bifurcating branch of periodic solutions we made the calculations necessary to show that the critical constant $\delta(T_0) > 0$ is positive (see Section 5.2 and Poore (1976)). Thus, as in the unharvested case studied in Section 5.2, we find that nonconstant p-periodic solutions of period near and larger than $p = 2\pi/y$ (which in this example is $p \sim 12.82$) for delays near and larger than the critical value T_0.

Recall that without harvesting $(H = 0)$ bifurcation occurs at $T = 2/b$ (Section 5.2). Thus, from FIGURE 7.1 we see that constant rate harvesting causes this bifurcation to occur earlier (for smaller delays T).

It would be of interest to see what effects delays have on other forced or control problems for other (multi-) species models (such as, for example, those nondelay problems considered in Svirezhev and Yelizarov (1972)).

REFERENCES

rbalat, I. 1959. Systémes d'équations differentielles d'oscillation non linéares, Rev. Math. pure et appl. 4, 267-270.

ddington, J. R. and May, R. M. 1975. Time delays are not necessarily destabilizing, Math. Biosci. 27, 109-117.

rsellino, A. and Torre, V. 1974. Limits to growth from Volterra theory of populations, Kybernetik 16, 113-118.

wnds, J. M. and Cushing, J. M. 1975. On the behavior of solutions of predator-prey equations with hereditary terms, Math. Biosci. 26, 41-54.

auer, F. 1977a. Stability of some population models with delay, Math. Biosci. 33, 345-358.

auer, F. 1977b. Asymptotic stability of a class of integro-differential equations, preprint.

peron, J. 1969. Time lag in population growth response of Isochrysis Galbana to a variable nitrate environment, Ecology 50, No. 2, 188-192.

swell, H. 1972. A simulation study of a time lag population model, J. Theor. Biol. 34, 419-439.

ppel, W. A. 1965. Stability and Asymptotic Behavior of Differential Equations, D. C. Heath and Co., Boston.

amer, N. F. and May, R. M. 1972. Interspecific competition, predation and species diversity: a comment, J. Theor. Biol. 34, 289-293.

shing, J. M. 1975. An operator equation and bounded solutions of integrodifferential systems, SIAM J. Math. Anal. 6, No. 3, 433-445.

shing, J. M. 1976a. Predator prey interactions with time delays, J. Math. Biol. 3, 369-380.

shing, J. M. 1976b. Stable limit cycles of time dependent multispecies interactions, Math. Biosci. 31, 259-273.

shing, J. M. 1976c. Periodic solutions of two species interaction models with lags, Math. Biosci. 31, 143-156.

shing, J. M. 1976d. Forced asymptotically periodic solutions of predator-prey systems with or without hereditary effects, SIAM J. Appl. Math. 30, No. 4, 665-674.

shing, J. M. 1976e. Periodic solutions of Volterra's population equation with hereditary effects, SIAM J. Appl. Math. 31, No. 2, 251-261.

hing, J. M. 1977a. Errata to "Periodic solutions of Volterra's population equation with hereditary effects", SIAM J. Appl. Math. 32, No. 4, 895.

shing, J. M. 1977b. Time delays in single species growth models, to appear in J. Math. Biol.

Cushing, J. M. 1977c. Bifurcation of periodic solutions of integrodifferential systems with applications to time delay models in population dynamics, SIAM J. Appl. Math. 33, No. 4.

Cushing, J. M. 1977d. On the oscillatory nature of solutions of general predator-prey models with time delays, Nonlinear Analysis (to appear).

Cushing, J. M. 1977e. Periodic time-dependent predator-prey systems, SIAM J. Appl. Math. 32, No. 1, 82-95.

Cushing, J. M. 1978. Stable positive periodic solutions of the time dependent logistic equation under possible hereditary influences, to appear in J. Math. Anal. Appl.

d'Ancona, U. 1954. The Struggle for Existence, Brill, Leiden.

Dunkel, G. M. 1968a. Some mathematical models for population growth with lags, thesis, University of Maryland, College Park, Md.

Dunkel, G. 1968b. Single species model for population growth depending on past history, in Seminar on Differential Equations and Dynamical Systems, Lec. Notes in Math. 60, Springer-Verlag, New York, 92-99.

Ehrlich, P. R. and Birch, L. C. 1967. The "balance of nature" and "population control", Am. Nat. 101, 97-107.

Gause, G. F. 1971. The Struggle for Existence, Dover, New York.

Gilpin, M. E. and Ayala, F. J. 1973. Global models of growth and competition, Proc. Nat. Acad. Sci. (U.S.A.) 70, 3590-3593.

Gomatam, J. and MacDonald, N. 1975. Time delays and stability of two competing species, Math. Biosci. 24, 247-255.

Gompertz, B. 1825. On the nature of the function expressive of the law of human mortality and on a new model of determining life contingencies, Phil. Trans. Roy. Soc. (London) 115, 513-585.

Hadeler, K. P. 1976. On the stability of the stationary state of a population growth equation with time-lag, J. Math. Biol. 3, 197-201.

Hairston, N. G., Smith, F. E., and Slobodkin, L. B. 1960. Community structure, population control, and competition, Am. Nat. 94, 421-425.

Halanay, A. 1966. Differential Equations, Stability, Oscillations, Time Lags, Monographs in Mathematics in Science and Engineering, Vol. 23, Academic Press, New York.

Halanay, A. and Yorke, J. A. 1971. Some new results and problems in the theory of differential-delay equations, SIAM Review 13, No. 1, 55-80.

Hale, J. 1971. Functional Differential Equations, Springer-Verlag series in Matt Sci. 3, Springer-Verlag, New York.

Holling, C. S. 1965. The functional response of predators to prey denisty and it role in mimicry and population regulation, Mem. Ent. Soc. Canada 45, 3-60.

nes, G. S. 1962a. The existence of periodic solutions of $f'(x) = -\alpha f(x - 1)\{1 + f(x)\}$, J. Math. Anal. Appl. 5, 435-450.

nes, G. S. 1962b. On the nonlinear differential-difference equation $f'(x) = -\alpha f(x - 1)\{1 + f(x)\}$, J. Math. Anal. Appl. 4, No. 3, 440-469.

kutani, S. and Markus, L. 1958. On the non-linear difference-differential equation $y'(t) = [A - By(t - \tau)]y(t)$, in Contributions to the Theory of Nonlinear Oscillations, Princeton Univ. Press, New Jersey, 1-18.

plan, J. and Yorke, J. A. 1975. On the stability of a periodic solution of a differential delay equation, SIAM J. Math. Anal. 6, No. 2, 268-282.

olle, H. 1976. Lotka-Volterra equations with time delay and periodic forcing term, Math. Biosci. 31, 351-375.

ch, A. L. 1974. Competitive coexistence of two predators utilizing the same prey under constant environmental conditions, J. Theor. Biol. 44, 387-395.

n, J. and Kahn, P. B. 1977. Limit cycles in random environments, SIAM J. Appl. Math. 32, No. 1, 260-291.

tka, A. J. 1956. Elements of Mathematical Biology, Dover, New York.

ckinbill, L. S. 1973. Coexistence in laboratory populations of Paramecium aurelia and its predator Didinium nasutum, Ecology 6, 1320-1327.

cArthur, R. H. and Levins, R. 1967. The limiting similarity, convergence, and divergence of coexisting species, Amer. Nat. 101, 377-385.

cDonald, N. 1976. Time delays in prey-predator models, Math. Biosci. 28, 321-330.

cDonald, N. 1977a. Time delays in prey-predator models: II. Bifurcation theory, Math. Biosci. 33,

cDonald, N. 1977b. Time lag in a model of a biochemical reaction sequence with end-product inhibition, preprint.

y, R. M. 1973. Time-delay versus stability in population models with two and three trophic levels, Ecology 54, No. 2, 315-325.

y, R. M. 1974. Stability and Complexity in Model Ecosystems, Monograph in Population Biology 6, Second edition, Princeton University Press, Princeton, N.J.

y, R. M., Conway, G. R., Hassell, M. P., and Southwood, T. R. E. 1974. Time delays, density-dependence and single species oscillations, J. Anim. Ecol. 43, No. 3, 747-770.

y, R. M. and Leonard, W. J. 1975. Nonlinear aspects of competition between three species, SIAM J. Appl. Math. 29, No. 2, 243-253.

zanov, A. and Tognetti, K. P. 1974. Taylor series expansion of delay differential equations - a warning, J. Theor. Biol. 46, 271-282.

ller, R. K. 1966. On Volterra's population equation, SIAM J. Appl. Math. 14, No. 3, 446-452.

Miller, R. K. 1971. Nonlinear Volterra Integral Equations, Benjamin Press, Menlo Park, California.

Miller, R. K. 1972. Asymptotic stability and perturbations for linear Volterra integrodifferential systems, in Delay and Functional Differential Equations and their Applications, edited by K. Schmitt, Academic Press, New York.

Miller, R. K. 1974. Structure of solutions of unstable linear Volterra integrodifferential equations, J. Diff. Eqns. 15, No. 1, 129-157.

Murdoch, W. W. 1966. "Community structure, population control, and competition"--a critique, Am. Nat. 100, 219-226.

Murray, J. D. 1976. Spatial structures in predator-prey communities - a nonlinear time delay diffusional model, Math. Biosci. 30, 73-85.

Nicholson, A. J. 1954. An outline of the dynamics of animal populations, Austr. J. Zool. 2, 9-65.

Nussbaum, R. D. 1973. Periodic solutions of some nonlinear, autonomous functional differential equations II, J. Diff. Eqns. 14, 360-394.

Nussbaum, R. D. 1974. A correction of "Periodic solutions of some nonlinear, autonomous functional differential equations II", J. Diff. Eqns. 16, 548-549.

Parrish, J. D. and Saila, S. B. 1970. Interspecific competition, predation and species diversity, J. Theor. Biol. 27, 207-220.

Pielou, E. C. 1969. An Introduction to Mathematical Ecology, Wiley-Interscience, New York.

Poole, R. W. 1974. An Introduction to Quantitative Ecology, Series in Population Biology, McGraw-Hill, New York.

Poore, A. B. 1976. On the theory and application of the Hopf-Friedrichs bifurcation theory, Arch. Rat. Mech. Anal. 60, No. 4, 371-393.

Rabinowitz, P. H. 1971. Some global results for nonlinear eigenvalue problems, J. Func. Anal. 7, 487-513.

Rescigno, A. 1968. The struggle for life: II, Three competitors, Bull. Math. Biophys. 30, 291-298.

Rescigno, A. and Richardson, I. W. 1973. The deterministic theory of population dynamics, in Foundations of Mathematical Biology, edited by Robert Rosen, Academic Press, New York, 283-359.

Ricklefs, R. E. 1973. Ecology, Chiron Press, Inc., Newton, Mass.

Rosenzweig, M. L. 1971. Paradox of enrichment: destabilization of exploitation ecosystems in ecological time, Science 171, 385-387.

Ross, G. G. 1972. A difference-differential model in population dynamics, J. Theor. Biol. 37, 477-492.

Scott, E. J. 1969. On a class of linear partial differential equations with retarded argument in time, Bul. Inst. Poli. Din Iazi, T. XV (XIX), 99-103.

:udo, F. M. 1971. Vito Volterra and theoretical ecology, Theor. Pop. Biol. 2, 1-23.

eifert, G. 1973. Liapunov-Razumikhin conditions for stability and boundedness of functional differential equations of Volterra type, J. Diff. Eqns. 14, 424-430.

lobodkin, L. B., Smith, F. E., and Hairston, N. G. 1967. Regulation in terrestrial ecosystems, and the implied balance of nature, Am. Nat. 101, 109-124.

nale, S. 1976. On the differential equations of species in competition, J. Math. Biol. 3, 5-7.

nith, F. E. 1954. Quantitative aspects of population growth, in Dynamics of Growth Processes, edited by E. J. Boell, Princeton Univ. Press, Princeton, N.J.

nith, F. E. 1963. Population dynamics in Daphnia Magna and a new model for population growth, Ecology 44, No. 4, 651-663.

nith, J. M. 1974. Models in Ecology, Cambridge Univ. Press, Cambridge.

:rebel, D. E. and Goel, N. S. 1973. On the isocline methods for analyzing prey-predator interactions, J. Theor. Biol. 39, 211-239.

virezhev, Yu. M. and Yelizarov, Ye. Ya. 1972. Mathematical Models of Biological Systems, Vol. 20 of Problems of Space Biology, NASA TT-780 (Translation, 1973), NASA, Washington, D. C.

rager, W. 1970. Symbiosis, Series on Selected Topics in Modern Biology, Van Nostrand Reinhold Co., New York.

erhulst, P. F. 1838. Notice sur la loi que la population suit dans son acroissement, Corr. Math. et Phys. 10, 113.

olterra, V. 1909. Sulle equazioni integro-differenziali della teoria dell' elasticitá, Rend. della R. Accad. dei Lincei 18, series 5, 2nd semestre, Vol. 9, 295-301.

olterra, V. 1927. Variazioni e fluttuazioni del numero d'individui in specie animali conviventi, R. Comit. TaLass. Ital., Memoria 131, Venezia.

olterra, V. 1931. Lecons sûr la théorie mathématique de la lutte pour la vie, Gauthiers-Villars, Paris.

alther, H. O. 1975a. Existence of a non-constant periodic solution of a non-linear autonomous functional differential equation representing the growth of a single species population, J. Math. Biol. 1, 227-240.

lther, H. O. 1975b. Stability of attractivity regions for autonomous functional differential equations, Manuscr. Math. 15, 349-363.

lther, H. O. 1976. On a transcendental equation in the stability analysis of a population growth model, J. Math. Biol. 3, 187-195.

ng, P. K. C. 1963. Asymptotic stability of a time-delayed diffusion system, Trans. ASME, J. Appl. Mech. 30, Series E, No. 4, 500-504.

Wang, P. K. C. and Bandy, M. L. 1963. Stability of distributed-parameter processes with time-delays, J. Electronics and Control 15, 343-362.

Wangersky, P. J. and Cunningham, J. W. 1956. On time lags in equations of growth, Proc. Nat. Acad. Sci. 42, 699-702.

Wangersky, P. J. and Cunningham, W. J. 1957a. Time lag in prey-predator population models, Ecology 38, No. 1, 136-139.

Wangersky, P. J. and Cunningham, W. J. 1957b. Time lag in population models, Cold Springs Harbor Symposium on Quantitative Biology 22, 329-337.

White, B. S. 1977. The effects of a rapidly-fluctuating random environment on systems of interacting species, SIAM J. Appl. Math. 32, No. 3, 666-693.

Wright, E. M. 1955. A non-linear difference-differential equation, J. Reine Angew. Math. 194, 66-87.

Editors: K. Krickeberg; S. Levin; R. C. Lewontin; J. Neyman; M. Schreiber

Biomathematics

Vol. 1: **Mathematical Topics in Population Genetics**
Edited by K. Kojima
55 figures. IX, 400 pages. 1970
ISBN 3-540-05054-X

This book is unique in bringing together in one volume many, if not most, of the mathematical theories of population genetics presented in the past which are still valid and some of the current mathematical investigations.

Vol. 2: E. Batschelet
Introduction to Mathematics for Life Scientists
200 figures. XIV, 495 pages. 1971
ISBN 3-540-05522-3

This book introduces the student of biology and medicine to such topics as sets, real and complex numbers, elementary functions, differential and integral calculus, differential equations, probability, matrices and vectors.

M. Iosifescu; P. Tautu
Stochastic Processes and Applications in Biology and Medicine

Vol. 3: Part 1: Theory
331 pages. 1973
ISBN 3-540-06270-X

Vol. 4: Part 2: Models
337 pages. 1973
ISBN 3-540-06271-8

Distribution Rights for the Socialist Countries: Romlibri, Bucharest

This two-volume treatise is intended as an introduction for mathematicians and biologists with a mathematical background to the study of stochastic processes and their applications in medicine and biology. It is both a textbook and a survey of the most recent developments in this field.

Vol. 5: A. Jacquard
The Genetic Structure of Populations
Translated by B. Charlesworth; D. Charlesworth
92 figures. Approx. 580 pages. 1974
ISBN 3-540-06329-3

Population genetics involves the application of genetic information to the problems of evolution. Since genetics models based on probability theory are not too remote from reality, the results of such modeling are relatively reliable and can make important contributions to research. This textbook was first published in French; the English edition has been revised with respect to its scientific content and instructional method.

**Springer-Verlag
Berlin
Heidelberg
New York**

This series aims to report new developments in biomathematics re search and teaching – quickly, informally and at a high level. The type of material considered for publication includes:

1. Preliminary drafts of original papers and monographs
2. Lectures on a new field, or presenting a new angle on a classical field
3. Seminar work-outs
4. Reports of meetings, provided they are
   a) of exceptional interest and
   b) devoted to a single topic.

Texts which are out of print but still in demand may also be considered if they fall within these categories.

The timeliness of a manuscript is more important than its form, which may be unfinished or tentative. Thus, in some instances, proofs may be merely outlined and results presented which have been or will later be published elsewhere. If possible, a subject index should be included. Publication of Lecture Notes is intended as a service to the internationa scientific community, in that a commercial publisher, Springer-Verlag can offer a wider distribution to documents which would otherwise have a restricted readership. Once published and copyrighted, they car be documented in the scientific literature.

Manuscripts

Manuscripts should comprise not less than 100 and preferably not more than 500 pages.
They are reproduced by a photographic process and therefore must be typed with extreme care. Symbol not on the typewriter should be inserted by hand in indelible black ink. Corrections to the typescrip should be made by pasting the amended text over the old one, or by obliterating errors with white cor recting fluid. Authors receive 75 free copies and are free to use the material in other publications. Th typescript is reduced slightly in size during reproduction; best results will not be obtained unless th text on any one page is kept within the overall limit of 18 x 26.5 cm (7 x 10½ inches). The publisher will be pleased to supply on request special stationery with the typing area outlined.

Manuscripts in English, German or French should be sent to Dr. Simon Levin, Center for Appli Mathematics, Olin Hall, Cornell University Ithaca, NY 14850/USA or directly to Springer-Verlag Heidelbe

Springer-Verlag, Heidelberger Platz 3, D-1000 Berlin 33
Springer-Verlag, Neuenheimer Landstraße 28–30, D-6900 Heidelberg 1
Springer-Verlag, 175 Fifth Avenue, New York, NY 10010/USA

ISBN 3-540-**08449**-5
ISBN 0-387-**08449**-5